THE NATURAL HISTORY ESSAYS

PEREGRINE SMITH LITERARY NATURALISTS

THE NATURAL HISTORY ESSAYS

HENRY DAVID THOREAU

Introduction and notes by
ROBERT SATTELMEYER

PEREGRINE SMITH BOOKS
SALT LAKE CITY

Printed in the United States of America
92 91 90 89 6 5 4 3 2 1

Library of Congress Cataloging in Publication Data

Thoreau, Henry David, 1817-1862.
Natural history essays.

(Peregrine Smith literary naturalists)
1. Natural history. 2. Natural history—Massachusetts.
I. Title. II. Series.
QH81.T6122 1988 917.44′02 88-34468
ISBN 0-87905-298-8

The paper used in this publication meets the minimum requirements of American National Standard for Information Sciences—Permanence of Paper for Printed Library Materials, ANSI Z39.48-1984. ∞

CONTENTS

INTRODUCTION

Ever since the first explorers sent back to Europe enthusiastic and distorted accounts of the natural wonders of the new continent, natural history writers have played a large role in defining the nature of American experience. The underlying mythology of the eras of exploration and settlement made the American an Adam-like figure, given a new world and and the opportunity to make himself and society over without the Old World's traditionary weaknesses. The naturalist's role was no less than a new version of Adam's charge in paradise: to name and describe each living thing man was to have dominion over. On a less mythological level, natural history writing provided Americans with an inventory of their riches and a forum for important debate about the relations of man to nature and about the nature of nature itself in the New World. Two of the finest works concerned with natural history in the eighteenth century, for example, William Bartram's *Travels* and Thomas Jefferson's *Notes on the State of Virginia*, used their subject matter to construct sophisticated visions of the character and potential of life in America — Bartram's purpose being to dramatize an Enlightenment Quaker's reasoned rapture at the works of God, and Jefferson's being to defend American nature (and by extension Americans themselves) from the prevailing European theory asserting the inferiority and degeneracy of natural products in the New World. Nineteenth and twentieth century classics such as John Wesley Powell's *Report on the Lands of the Arid Regions of the United States*, Aldo Leopold's *Sand County Almanac*, and Rachel Carson's *Silent Spring* contained powerful critiques of some fundamental American assumptions about the uses to which nature may be put, and have had a significant impact on the shaping of laws and public policy. The sheer fact of nature itself — its overwhelming presence, its difference from familiar European norms, its seemingly limitless extent — has always been a major component of national self-definition. The naturalist's challenge has been not merely to describe this massive and protean phenomenon but to interpret its significance for civilization and ultimately render it meaningful to human consciousness.

Yet the naturalist has also been a comic or even a suspicious figure in America. Dr. Battius in Cooper's *The Prairie* was a caricature of the early naturalist, absently endangering himself and others in an addled quest for new species, and spouting unintelligible Latin phrases from the new Linnaean taxonomic system. At a deeper level, though, the naturalist's pursuits could even be subversive of the social and economic order. In a pragmatic, expanding society whose major enterprises were clearing, settling, farming, and building, the destruction of habitat and the displacement if not the extinction of native species was inevitable. In this environment the naturalist was doubly suspect because he was concerned about wild plants and animals, and — perhaps even more disturbingly — because he seemed to do no work. His studies necessitated the kind of patient observation of often minute phenomena which could only seem trivial to the mass of his contemporaries.

Although the prevalent atmosphere of Concord, Massachusetts, was probably more tolerant than that of most places in mid-nineteenth century America, Henry Thoreau came in for his share of this distrust and antagonism. He even wrote *Walden*, in part, to answer his neighbors' inquiries about what seemed to them an idle way of life for a young man whose family had sacrificed to send him to Harvard College. And in "Life Without Principle," an essay which dealt in a more concentrated way with the problem of making a living without losing one's soul, he put the problem succinctly: "If a man walk in the woods for love of them half of each day, he is in danger of being regarded as a loafer; but if he spends his whole day as a speculator, shearing off those woods and making earth bald before her time, he is esteemed an industrious and enterprising citizen."

Americans in the late twentieth century, who work very hard at cultivating leisure, are probably more willing than Thoreau's contemporaries were to grant him the broad margin of free time that he required; and his advocacy of the wild as a necessary complement to civilized life has made him something of an environmental oracle, whose words embellish countless calendars, posters, and collections of photographs. Moreover, in a curious turn of events for a writer who was largely ignored in his own time, his work has become a mark against which later writers are inevitably judged and usually found lacking: it is almost as fatal for a naturalist with literary pretensions to be compared to Thoreau as it is for a humorist to be compared to Mark Twain.

Still, this popularity and preeminence is largely based on just a few episodes in Thoreau's life — his two-year stay at Walden Pond and his one-night stay in jail — and on the widespread familiarity of certain sentences from *Walden* and the essay "Walking": "The mass of men lead lives of quiet desperation," for example, or "In Wildness is the preservation of the World." The actual records of his lifelong attention to natural history — his voluminous Journal, his travel books *Cape Cod* and *The Maine Woods*, and the essays included in this volume — are comparatively little studied. In fact, Thoreau's natural history writing and his abilities as a naturalist have frequently even been denigrated, and the concensus of critical opinion on this aspect of his career is a puzzling paradox, most clearly evident in Sherman Paul's pronouncement in *The Shores of America* that "In spite of his gifts for nature study, Thoreau was not a good naturalist . . . " (p. 277). How and why this odd state of affairs came to be can only be understood in the context of Thoreau's development as a natural historian and of the popular understanding of the mission and methods of science in his day and ours.

The foundation of Thoreau's interest in natural history was his passionate affection for his native environment. Concord was a rural village, and Thoreau's boyhood pursuits were those of most of his fellows — hunting and fishing, boating, berry-picking — augmented by his family's cultivation of outdoor activities. With his older brother John, especially, he ranged the countryside, collecting Indian artifacts and watching birds. The two built their own boat, and in 1839 they took it on a two-week trip down the Concord and up the Merrimack Rivers, after which they hiked overland to Mt. Washington in the White mountains. Years later, after John's sudden death in 1842, Thoreau would commemorate the journey in his first book, *A Week on the Concord and Merrimack Rivers*. In a Harvard classbook entry, Thoreau described his youth in Wordsworthian terms, saying: "Those hours that should have been devoted to study, have been spent in scouring the woods, and exploring the lakes and streams of my native village." And as a young man he so impressed Emerson and Hawthorne with his boating skill and knowledge of riparian life that both wrote enthusiastic accounts in their journals of excursions in his company on Concord streams.

By the philosophical currents of his age and his own intellectual training, however, Thoreau looked on nature as much more than the source of recreation or picturesque scenery: it was the phenomenal medium through which divinity and truth were communicated to man. Even more insistently than other nineteenth century philosophical and religious movements, the Transcendentalists were preoccupied with the great question of the age — in the words of *Nature,* Emerson's important manifesto of the movement, "to what end is nature?" Although the notion that nature provided a link with divinity might seem to ally Thoreau in spirit to earlier naturalists like William Bartram or Gilbert White, who saw in nature the wonderful harmony and balance of God's creation revealed, a crucial difference existed. To the earlier naturalists (and to many of their descendants today) nature was the *evidence* of design in the universe; by studying the details and organization of nature one could discover its universal laws and even infer the attributes of its creator, in much the same way that anthropologists might try to reconstruct the culture of a prehistoric people by studying its fossil artifacts. Natural history was a tool, frequently, of what was called Natural Theology, where students expected to find, as the title of a popular book on geology expressed it, *The Foot-Prints of the Creator.*

Thoreau, like many other Romantics, was intensely interested in science, but he was less interested in the footprints of the Creator than he was in creation itself. He was disposed to find in nature not the result of some previous plan but a phenomenon continually expressive of creation; not the evidence of design but design itself. Nature was the medium through which spirit manifested itself, and Thoreau was quite in earnest when he proclaimed in the climactic "Spring" chapter of *Walden* that "The earth is not a mere fragment of dead history, stratum upon stratum like the leaves of a book, to be studied by geologists and antiquaries chiefly, but living poetry like the leaves of a tree, which precede flowers and fruit, — not a fossil earth, but a living earth."

The essays in which Thoreau reported his investigation of nature and the natural history of New England fall naturally into two groups. The first are apprentice and exploratory pieces growing out of his earliest engagement with nature and the writer's craft. The essays in the second group were quarried from lectures and unfinished longer studies near the end of his life, and represent that portion of his late work he was able to put in shape for publication before his death in 1862. The reader interested in the overall development of Thoreau's natural history should bear in mind that in between these two distinct groups of essays lie Thoreau's two books — *A Week on the Concord and Merrimack Rivers* and *Walden* — and his book-length collections of travel and natural history essays — *The Maine Woods* and *Cape Cod.*

After Thoreau graduated from Harvard in 1837 he thought of himself (during the hours he could spare from schoolteaching) as an apprentice man of letters, and at first he tried the conventional literary forms of criticism, poetry, and the familiar essay. His efforts in these fields, however, met with little success, even from the sympathetic editors and readers of the Transcendentalists' magazine, *The*

Dial, where they were published. His mentor Emerson saw more clearly than Thoreau at this time the bent of his young friend's genius, and one of his first acts upon assuming the editorship of *The Dial* in 1842 was to ask Thoreau to review several volumes of surveys of the flora and fauna of Massachusetts that had recently been commissioned and published by the state. He explained to Thoreau "the felicity of the subject to him, as it admits of the narrative of all his woodcraft boatcraft & fishcraft."

Thoreau accepted the commission, and produced for the July 1842 issue "Natural History of Massachusetts," his first essay on the subject which was increasingly to absorb his energies for the rest of his career. "Natural History of Massachusetts" is patently an apprentice work which neither finds a structure of its own nor actually reviews the volumes named. But Thoreau blithely admits to having placed the surveys at the beginning of the essay "with as much license as the preacher selects his text," and the interest and energy of the ensuing discourse derive chiefly from self-discovery. Although Thoreau begins by stressing, scholar-like, the winter activity of reading about natural history, this imposed distance from his subject quickly disappears as his rambling catalogue of New England sights and scenes begins. Most important for his future as a naturalist and as a writer, however, is his concluding formulation about the nature of natural history, a passage which really amounts to a kind of prologomena to his life's work, and which sets forth the challenge facing the Transcendentalist as would-be scientist:

> "The true man of science will know nature better by his finer organization; he will smell, taste, see, hear, feel, better than other men. His will be a deeper and finer experience. We do not learn by inference and deduction and the application of mathematics to philosophy, but by direct intercourse and sympathy. It is with science as with ethics, — we cannot know truth by contrivance and method; the Baconian is as false as any other, and with all the helps of machinery and the arts, the most scientific will still be the healthiest and friendliest man, and possess a more perfect Indian wisdom."

Behind the tone of youthful self-assurance in this passage lies a crucial, though perhaps not an apparent, distinction. When Thoreau asserts that "the Baconian is as false as any other," he may seem to be dismissing casually the whole revered inductive superstructure of what we have come to think of as the Scientific Method, in favor of some undefined and possibly mystical "Indian wisdom." But what he actually says, of course, is that "the Baconian *is as false as any other*." This is a statement of principle — a fundamental assumption — that no theory of nature or way of representing nature should be mistaken for nature itself. No result produced by "contrivance and method," no matter how attractive or useful, should be confused with nature's essence.

The challenge Thoreau set for himself at the beginning of his career, one he would respect if not always be comfortable with for the rest of his life, was how to contribute legitimately to the natural history of New England without succumbing to the lure of method and material results.

Thoreau was encouraged enough by this first effort to turn almost immediately to similar projects. He wrote "A Walk to Wachusett," based on a trip he made in July of 1842, during the following fall. He began "A Winter Walk" at the same time, although he did not complete the essay until June 1843. The sophistication of his natural history did not increase significantly in these essays, but he did mature greatly as a writer and begin to master the skills which would serve as the basis of his later work. He made strides to correct the anecdotal and rambling character of "Natural History of Massachusetts" in "A Walk to Wachusett" by employing a narrative rather than discursive form, and by subordinating the mere observation of nature to an archetypal pattern — the quest — which deepened the significance of what was observed. The specific quest of "A Walk to Wachusett," to see if nature is capable of sustaining the imaginative significance with which the narrator invests it, has perhaps a foregone conclusion. But the pattern enabled Thoreau to place his observation of nature in its true context of spiritual aspiration. Henceforth the universal and timeless patterns of the journey out and back, and the ascent and descent of mountains, together with the diurnal and seasonal cycles, would provide him with both structural principles and a symbolic dimension for his writing about nature.

Thoreau still tended to see nature, however, through European conventions of landscape description. "A Walk to Wachusett" focuses less on specific natural detail than on sweeping and sometimes painterly descriptions of broad panoramas or of the picturesque occupations — cultivating hops, for instance — of the country people. "A Winter Walk," on the other hand, is much closer in tone and spirit to his mature work, for in it he began to concentrate on the particular and even minute details of his environment, and to make this newly-sharpened perception into a kind of tacit source of value, through a central image or conceit running through the whole essay: the idea of feeling summer warmth in winter.

Emerson almost rejected "A Winter Walk" for *The Dial* because of this theme of finding warmth in cold, which he considered to be no more than a perverse mannerism of Thoreau's, a symptom of his unfortunate fondness for paradox. But this motif is absolutely essential to the charge of meaning with which Thoreau wishes to endow the process of sense perception. Nature is cold and dead this winter morning, until the narrator's imagination, fueled by the flow of sense perception, begins to generate an "increased glow of thought and feeling." If there is a "slumbering subterranean fire in nature" which not even the intensest cold can extinguish, it is because this warmth answers to the "subterranean fire . . . in each man's breast." The reciprocity of nature and man's imagination produces the warmth by which life is maintained. Characteristically, Thoreau finds an apparent paradox to be his most effective way of expressing this truth, for it is always the inside of the outside he seeks to reveal, and "our vision does not penetrate the surface of things," as he put it in *Walden*.

A corollary of this emphasis on clarity of perception is his discovery — analogous to Faulkner's discovery of an epic world in his poor northern Mississippi county — that the smallest details of nature may tell the most important stories. Many years later in the introduction to his lecture on "Huckleberries" he would quote Pliny approvingly: *In minimis Natura praestat* — Nature excels in the least things. He began to pay attention to this important truth in "A Winter Walk," noting in almost microscopic detail the "submarine cottages of the caddice-worms," the "tiny tracks of mice around every stem," and the chip of wood which "contains inscribed on it the whole history of the woodchopper and his world."

Yet, on balance, these early essays testify that Thoreau was still an admirer of nature whose enthusiasm and gifted amateur eye masked the fact that he actually knew relatively little systematic natural history. Additionally, the familiar essay was less than ideally suited to his talents at this stage in his life. In the process of composition his imagination worked slowly and accretively, normally requiring years to raise the structures of his works, and the quickly turned out magazine piece was rarely a suitable form for him. Both his education and his craft required long seed times.

The next decade of his life was devoted to the experiment at Walden Pond (1845-1847), to the writing of *A Week on the Concord and Merrimack Rivers* (1849) and *Walden* (1854), and to the acquisition of something approaching a professional competence as a naturalist. He began to botanize systematically, acquired a basic library of botanical guides, and learned taxonomy. He corresponded with and collected specimens for Louis Agassiz of Harvard, America's leading scientist. He also knew the work of Asa Gray, Harvard's other eminent natural scientist, destined to become Agassiz's rival in the American debate over Darwin's theory of evolution. He made a study of limnology and of the fishes of Concord rivers and ponds. He read Kirby and Spence and others on insects, and struck up a professional acquaintance with Thaddeus Harris, a prominent entomologist and the librarian of Harvard College. He had been interested in ornithology since boyhood (a family album of bird sightings survives, dating from the 1830s and containing entries by Henry, his brother John, and his sister Sophia), and he compiled a large collection of birds' nests and eggs. In 1850 he was elected a corresponding member of the Boston Society of Natural History, to which he contributed specimens and various written accounts over the years, and whose library and collections he used regularly in pursuing his studies. (His own extensive collections of Indian artifacts, birds' nests and eggs, and pressed plants went to the Society after his death.) In his work as a surveyor he made a more intimate acquaintance with the farms, swamps, and woodlots of Concord. Gradually his townsmen, who had generally looked askance at his activities, began to come around for help in identifying plants and animals, and to bring him new items for his collections.

In addition to his more systematic reading and collecting, the most important factors in his growth as a careful observer were his daily stints of walking and journal writing. As he states in "Walking," he averaged at least four hours a day in the field in all weathers, after which he carefully recreated from records kept in pocket notebooks the accounts of his excursions which began to swell the Journal. The bulk of this systematic observation came after 1850, so that thirteen of the fourteen published volumes of the Journal cover only the last eleven years of his active life, from 1850 to 1861.

The vast quantity of observation and raw data in the late Journal has led most critics to see in Thoreau's increasing attention to the collection of facts a loss of creative power. But this conclusion is based on a kind of false statistic, for there is still about the same amount of reflection and contemplation of ideas in the late Journal as the early Journal; it only appears to be more scattered because

Thoreau now used the Journal for the additional purpose of making detailed records of his various natural history observations. In the same way commentators point to Thoreau's occasional moments of concern or despondency over becoming too absorbed in detail, ignoring the larger fact that he worked happily and with increasing energy on these studies until his final illness, when he prepared as many of his papers as possible for publication.

At any rate, the results of his new diligence and competence as a naturalist were manifest in his writings — the "Concord River" chapter of *A Week*, the exquisite miniatures as well as the larger studies of the pond itself in *Walden*, and in his accounts of the northern wilderness and the wild seashore in the chapters of *The Maine Woods* and *Cape Cod* that were published serially in the 1840s and 1850s. And as far as Thoreau himself was concerned, the real danger to his career lay not in becoming too scientific, but in becoming estranged from the scientific community itself. By continuing in the midst of his detailed studies to hold to a vision of a humane science which would not treat nature merely as matter to be manipulated, he realized that he was stemming an ever-strengthening tide of belief to the contrary. When he was proposed for membership in the Association for the Advancement of Science in 1853, he was asked to complete a form describing his particular field of study, and he realized to his dismay what would happen if he told the truth: "Now, though I could state to a select few that department of human inquiry which engages me, and should be rejoiced at an opportunity to do so, I felt that it would be to make myself the laughing-stock of the scientific community to describe or attempt to describe to them that branch of science which specially interests me, inasmuch as they do not believe in a science which deals with the higher law."

Nevertheless, Thoreau did attempt to explain himself, usually by lecturing to local lyceums in New England. All his later natural history essays, in fact, began as lectures, and he did not put them into essay form until near the end of his life. His most concerted attempt to set forth the rationale of his way of life and to explain why civilization could not afford to cut itself off from its wild heritage is "Walking," an essay which he spliced together from two lectures, "Walking" and "The Wild," which he gave many times during the 1850s.

"Walking" is one of Thoreau's better-known essays today because its advocacy of the wild is the philosophical cornerstone of twentieth century movements to preserve wilderness tracts in America. Yet Thoreau did not think of the wild — or of walking either — as a special preserve. It was not for recreation so much as it was for re-creation; it was a particular quality of life that had to be actively cultivated. The spirit of the walk and not the specific route makes one a true saunterer (*Sainte-Terrer*, Holy-Lander, in his not entirely fanciful etymology), because the walk undertaken rightly denotes a commitment to the highest uses to which thought and observation may be put. Even Thoreau's paean to the West — jingoistic as it may appear at first — is primarily in praise of man's capacity to imagine and live according to his vision of a fairer world. "Westward I go free" may have the ring of pioneer travel about it, but Thoreau had to travel no farther than the Old Marlborough Road in Concord to find his West.

Similarly, the wild has less to do with actual wilderness areas (for that Thoreau went to Maine) than it does with a habit of mind which recognizes the balance of mutually dependent forces in life. The avowed "extreme statement" on behalf of the wild is not atavism or even primitivism, but an attempt to redress an imbalance in our way of thinking about life in nature. The wild is a reminder of an original attachment to the sources of life, and points back to a time and a state where nature and man's consciousness were not separate entities, and where nature was not an object to be learned and mastered for the sake of material knowledge and power. Hence Thoreau's proposal, at once earnest and ironic, for a "Society for the Diffusion of Useful Ignorance," for he argues that it is only when we become wise enough to forget that we "know" nature that we can participate in what he calls "Beautiful Knowledge." The historical result of increasing knowledge in the scientific sense has been an increased separation and loss of harmony between man and nature, a loss, as Thoreau points out, that can be demonstrated by history and language itself: "We have to be told that the Greeks called the world **Κοσμος**, Beauty, or Order, but we do not see clearly why they did so, and we esteem it at best only a curious philological fact." The wild is valuable and necessary insofar as it enables us to imagine, however fleetingly, that lost harmony.

This belief that nature, if viewed from the correct perspective, could provide one with a way of realigning himself with the sources of beauty and harmony was based on close and careful investigation, and was the result of hard work as much as inspiration. As the 1850s progressed, Thoreau focused increasingly on observing and recording the yearly natural cycle of his local environment, with particular emphasis on the patterns of leafing and flowering, fruiting, and seed dispersal in plants. One offshoot of these larger studies was a discovery he outlined in "The Succession of Forest Trees," his most sustained treatment of what he termed a "purely scientific subject." In it he describes a phase in the evolution of a climax forest, dispelling still widely-held beliefs that trees were propagated by spontaneous generation or by seeds that lay dormant in the ground for many years. He also demonstrates that a naturalist who believes in "a science which deals with the higher law" can also produce accurate and useful insights into the operations of nature — and does so with a kind of self-deprecating humor that acknowledges his "outside" position: "Every man is entitled to come to a Cattle-show, even a transcendentalist."

Even in this discourse, however, meant to be informative and useful to his audience of local farmers, Thoreau's habitual perspective is evident. Although he insists that trees do not spring up by spontaneous generation or some other mysterious process, he dispels this myth in order to call attention to a more fundamental and real mystery — the seed: "Convince me that you have a seed there, and I am prepared to accept wonders." A case in point is the marvelous 186¼ pound squash — "*Poitrine jaune grosse*" — he raised in his garden: "These seeds were the bait I used to catch it, my ferrets which I sent into its burrow, my brace of terriers which unearthed it." Here, having just taken pains to disprove the popular fallacies about the generation of plants and to prove that they spring from seeds, Thoreau suddenly shifts the level of his argument to imply that the seed is not only related to the mature fruit by material cause and effect but is perhaps an organic principle in itself of yet another order.

This sort of intermingling of "poetic" and "scientific" truth has led most twentieth century critics to conclude that Thoreau was finally a poor naturalist on the one hand, and, after *Walden*, a failed creative artist as well, because he could not keep the elementary distinction between the two realms clear. But it seems odd that Thoreau, who was after all thoroughly grounded in the objective natural science of his day, should be so confused about such a fundamental point and find himself finally adrift somewhere between science and mysticism. Another possibility is that our own implicit and unexamined assumption about the unbridgeable gap between scientific and imaginative truth simply makes it almost impossible to grasp the nature of his work from the inside. Its elements appear to be anomalous because contemporary thought is unconsciously the intellectual heir of that Association for the Advancement of Science, before whom Thoreau was already unlikely to receive an impartial hearing; it being no less true in science than in war that victors write the histories.

At any rate, enough prejudice, conscious or unconscious, has existed towards Thoreau's kind of natural history that the bulk of his late work in this field has not been considered important enough to be published yet. Toward the end of his life he worked on two long manuscripts, one on seeds and the other on fruits, which the progressive debilitation of tuberculosis did not permit him to complete, but which do survive in preliminary draft versions (in the Berg Collection of the New York Public Library). Fortunately, these manuscripts are at last scheduled for publication, in the Princeton Edition of Thoreau's *Writings* now in progress, but they have not yet been fully considered in the central debate over Thoreau's career — whether his late years form a record of declining power and a straying from the vision that led to *Walden*, or whether they furnish evidence of significant new directions and works which he did not live to complete.

Some necessarily tentative and provisional proposals, however, about the direction of this late work might be advanced on the basis of published works which were collateral with or part of these longer projects: two essays Thoreau was able to compile from lecture drafts before he died, "Autumnal Tints" and "Wild Apples," and a small portion of the manuscript on fruits, called "Huckleberries," edited by Professor Leo Stoller, which is made widely available for the first time in this volume. Although the circumstances of their composition and publication suggest that they may have undergone further refinement at Thoreau's hands, and although they form only a small portion of much longer and more ambitious projects, these works at least give some hints about the vision and the program of natural science toward which Thoreau was working.

"Autumnal Tints" treats the leaf as fruit, and displays the concern with ripeness that dominates the imagery of Thoreau's late work. At its basic level, of course, the essay is a catalogue of the different leaves and leaf-tints of a New England fall, bearing witness to Thoreau's long-standing interest in this phenomenon. One of his earliest literary projects, back in 1841, had been a work called "The Fall of the Leaf." Although the subject obviously admits of a popular treatment (Thoreau's essay is, after all, a kind of literary precursor to the fall foliage tour), the leaf held a high and very special place among nature's forms to Thoreau. He regarded it, in fact, as the archetypal organic form, a kind of ur-phenomenon expressive of creative life. In the "Spring" chapter of *Walden* it is the narrator's climactic meditation on the leaf-forms expressed in the flowing sand of the railroad cut which signals his discovery of a vital principle in nature: "The Maker of this earth but patented a leaf." The leaf is a sort of universal hieroglyph or symbol of creative energy, and thus its ripening and its fall are events to be attended to with care. These events take on an even greater significance when it is recalled that the essay was prepared by Thoreau on his death bed, and that his treatment of the subject closely reflects his own condition.

One expression of the law to be discerned in the fall foliage is that "Generally, every fruit, on ripening, and just before it falls, when it commences a more independent and individual existence . . . acquires a bright tint. So do leaves," The high color is a sign of ripeness, not decay, and the fall itself is a sort of individuation, the commencement of "a more independent and individual existence," and not a death. This notion is of course at odds with the scientific explanation of what happens when a leaf falls (which Thoreau knew perfectly well), but his slant on natural facts, as should be evident by now, is deeply opposed in principle to the customary assumptions about what constitutes organic life or even reality. He is "more interested in the rosy cheek" than in "the particular diet the maiden fed on"; which is to say that while Thoreau is cognizant of the physiologist's explanation, he knows that the phenomenon is greater than the sum of these parts, and has no use for an explanation which fails to take into account the perception which shapes the appearance itself.

This much is clear from the concluding portion of the essay, where Thoreau refers more than once to the "intention of the eye" as a determinant of reality. This principle, which Thoreau derives and illustrates from experience, means in essence that one must know what he is looking for before he can see it. "The astronomer," as he says, "knows where to go star-gathering, and sees one clearly in his mind before any have seen it with a glass." The history of science itself suggests the basic reasonableness of this proposition (think, for example, of the theoretical anticipation of the major discoveries of physics in this century). But the proposition undercuts at the same time the cherished and popular myth of a perfectly objective and quantifiable world which presupposes nature as matter independent of and anterior to any perceiving consciousness.

In effect, Thoreau found himself fighting a kind of rear-guard or guerilla action against scientific materialism, and he adopted as his favored rhetorical strategies "extreme statement," paradox, and the deliberate inversion of accepted wisdom, in order to try to startle his audience out of unexamined and merely habitual modes of thought. In the introduction to "Huckleberries," for example, the customary standards of littleness and greatness are reversed, not in order to allow Thoreau to deliver a few broadsides at politics and education but to try to induce in his auditors a new perspective on familiar objects and ideas. Thoreau's natural history would ultimately undermine the social as well as the philosophical conventions, for the humble huckleberry is the launching point for a radical critique — almost reminiscent of Marx at times — of entrepreneurial activity which robs the community of its natural birthright and promotes the division of labor and the alienation of the worker from his work. Elsewhere in "Huckleberries" he sounds what by now has become a familiar plea for cities and towns to set aside a portion of their wild and uninhabited lands as a resource for future generations. In his own country, though, Thoreau was a prophet without honor: even Emerson was unable at last to follow sympathetically or to grasp the nature of his work, and in regard to "Huckleberries," this long-buried and unknown work, there is a kind of consummate irony in the famous criticism of Thoreau leveled by Emerson in his funeral oration: "instead of engineering for all America, he was the captain of a huckleberry party."

"Wild Apples" is the most complete and most compelling of these late works, and furnishes perhaps the best evidence of the range of reference of Thoreau's detailed studies of his native ground. And portions of the essay are memorable in their own right for narrative and descriptive verve—the struggle of the apple tree against its bovine foes and its eventual triumph over them, for instance — independent of any historical or scientific context. The subject was perfectly suited to Thoreau. The wild apple was appealing to him because it was a forgotten and neglected fruit, one of the "least things" at which nature excelled, flourishing in the unfrequented corners of New England its chronicler instinctively sought; and because its situation and its qualities were so transparently suggestive of his own: a cultivated plant tending back to the wild, bearing its fruit late and unnoticed by most, crabbed and gnarled perhaps, but bracing if taken in the right spirit.

Yet it finally requires the sort of altered perspective on natural facts which Thoreau strove to induce in his audiences if "Wild Apples" is to be seen as a coherent whole, for without such a perspective (or the willingness, at least, to entertain it) crucial parts of the essay will seem at best to be unrelated to the descriptive body of the piece. What, for example, is the function of the very detailed philological and historical account of the apple which begins the essay? It is far too detailed merely to "introduce" the subject; and since "Wild Apples" was originally a lecture which Thoreau revised for publication during his final illness, it seems equally unlikely that this long recital of definitions and historical facts is mere padding.

The key to this section is the deceptively bland opening sentence: "It is remarkable how closely the history of the Apple-tree is connected with that of man." Thoreau does more here than establish the groundwork for a metaphorical connection between man and the apple tree. He suggests, rather, by tracing the significance of the apple in language and history, that it is a natural fact which can best be understood through an understanding of its evolution in human thought. By giving as his opening coordinates, so to speak, not merely the genus and species of the apple, but its meaning in history, poetry, mythology, religion, and folklore, Thoreau suggests an alternative approach to customary scientific description, which involves actively putting the history back in natural history. Since the reality of natural phenomena is in part dependent upon the perceiver, true natural history involves the historical evolution of this perception. What the apple tree means, finally, is the sum of its histories, of its relationship to man. Man and the apple tree have grown up together.

The frame Thoreau provides for his subject, then, in this introduction, seeks to establish an interconnectedness between supposedly independent and discrete phenomena and human thought. He deliberately reverses the path of normal science, which seeks for objectivity to isolate and separate the object studied.

This historical dimension opens even more far-reaching levels of significance. The word for apple, we learn, if traced back far enough once meant "riches in general," all the productions of nature which were at man's disposal. And, as Thoreau also points out in an understatement of epic proportions, "Some have thought that the first human pair were tempted by its fruit." Thus the apple embodies simultaneously man's dominion over nature and his fall from harmony with it — the ultimate paradox of human knowledge. If we understand the story of the Fall at one level to represent the separation of man from an original harmony with nature, a falling into self-consciousness in which nature began to be perceived as *other*, then the history of the apple tree becomes the history of man's relation to nature. How the apple tree is defined and perceived is at any moment an index of our condition with respect to that original harmony and our prospects for regaining it.

Hence Thoreau's glorification of the wild apple, on the one hand, and his startling jeremiad at the end of the essay over its disappearance. He celebrates the wild apple because it suggests the possibilities of reattachment and harmony with nature without the sacrifice of knowledge. The apple tree has grown cultivated and domestic with man, and now aspires back to its original state without giving up its fruitfulness. That fruitfulness is all the more valuable because it is achieved after such a struggle. It suggests, in its balance of the wild and the cultivated, the possibility of victory over both ignorance and the tyranny of knowledge.

In this light the conclusion of the essay is a powerful and pertinent warning, and not merely a curious lapse into preaching, for it constitutes Thoreau's final plea against losing once and for all the possibility of achieving the harmony the apple tree suggests. "The era of the Wild Apple will soon be past," given the spread of those assumptions and habits of mind about nature which made his voice more and more isolated even in his own day. The inevitable result will be a universal malaise arising from the final separation of man from nature. Thoreau's choice of biblical text is hauntingly appropriate to his plea, for it is the death of nature and the alienation of man which ensue from the denial of creative spirit working through both: " 'The vine is dried up, and the fig-tree languisheth; the pomegranate-tree, the palm-tree also, and the apple-tree, even all the trees of the field, are withered: because joy is withered away from the sons of man.' "

The naturalist's mission as Thoreau finally expressed it went far beyond the naming of the products of the New World garden. Up through the writing of *Walden* his imagination had centered on the Spring as an emblem of physical and spiritual rebirth, but in his later years he became more concerned with the Fall as season and as spiritual fact. If that Fall was to be a fortunate one, our way of knowing must lead us back to and not away from its great central life.

Robert Sattelmeyer

A NOTE ON THE TEXT

With the exception of "Huckleberries," the essays in this volume originally appeared in periodicals during or just after Thoreau's lifetime: "Natural History of Massachusetts" and "A Winter Walk" in *The Dial* for July 1842 and October 1843, respectively; "A Walk to Wachusett" in the *Boston Miscellany* in January 1843; "Walking," "Autumnal Tints," and "Wild Apples" in the June, October and November 1862 issues of the *Atlantic Monthly*, respectively; and "The Succession of Forest Trees" in the *Transactions of the Middlesex Agricultural Society* for 1860. They were then collected in the first posthumous volume of Thoreau's works, *Excursions* (1863), which was edited by Sophia Thoreau and Ellery Channing. A work by Thoreau called "Night and Moonlight" was also included in the 1863 *Excursions*, but recent scholarship has shown it to be non-authorial in its printed form, and it has not been included in the present volume. (See William L. Howarth, "Successor to *Walden?* Thoreau's 'Moonlight — An Intended Course of Lectures,'" *Proof*, 2 [1972], pp. 89-115.) The text in this collection is a reprinting of the standard 1906 Walden Edition of *Excursions*. A definitive scholarly edition of *Excursions* is in preparation for *The Writings of Henry D. Thoreau*, in progress at Princeton University Press.

"Huckleberries" was not prepared for publication by Thoreau, but exists in an intermediate draft form in a longer manuscript labelled "Notes on Fruits" in the New York Public Library's Berg Collection of English and American Literature. The present text was edited, with textual notes and an introduction, by the late Leo Stoller, as *Huckleberries* (The Windhover Press of the University of Iowa and The New York Public Library, 1970). A scholarly text will appear when the "Notes on Fruits" manuscript is edited for a forthcoming volume of the Princeton Edition of Thoreau's *Writings*.

BIBLIOGRAPHICAL NOTE

Thoreau's interest in nature has of course received a great deal of attention, even if his natural history writings have been relatively little studied. The following is a highly selective list of works on both Thoreau's natural history and on the development of natural history in America.

Consideration of Thoreau's attitude toward science and the practice of natural history include Raymond Adams, "Thoreau's Science," *Scientific Monthly*, 60 (1945), 379-382; Nina Baym, "Thoreau's View of Science," *Journal of the History of Ideas*, 26 (1965), 221-234; Leo Marx evaluated "Thoreau's Excursions" in the *Yale Review*, 51 (1962), 363-369; a full-length study is James McIntosh, *Thoreau as Romantic Naturalist* (Ithaca: Cornell University Press, 1974); a provocative study of the intellectual and philosophical tradition within which Thoreau was working (although not specifically concerned with him) is Owen Barfield, *What Coleridge Thought* (Middletown, Conn.: Wesleyan University Press, 1971); the most comprehensive account of the development of Thoreau's thought is Sherman Paul, *The Shores of America* (Urbana: University of Illinois Press, 1958); and the most detailed biography is Walter Harding, *The Days of Henry Thoreau* (New York: Alfred A. Knopf, 1965).

Scholarly studies of natural history in America, including its treatment in literary works, include Philip Marshal Hicks, *The Development of the Natural History Essay in American Literature* (Philadelphia, 1924); Norman Foerster, *Nature in American Literature* (New York: The Macmillan Company, 1923); Elizabeth Sewell, *Orphic Voice: Poetry and Natural History* (New Haven: Yale University Press, 1960); Thomas Smallwood, *Natural History and the American Mind* (New York: Columbia University Press, 1941); Roderick Nash, *Wilderness and the American Mind*, rev. ed. (New Haven: Yale University Press, 1973).

Some recent popular accounts of early naturalists in America are Robert Elman, *First in the Field: America's Pioneering Naturalists* (New York: Mason/Charter, 1977); Wayne Henley, *Natural History in America* (New York: Quadrangle/The New York Times Book Company, 1977); and Joseph Kastner, *A Species of Eternity* (New York: Knopf, 1978).

This engraving of a Scarlet Oak leaf, which appeared in the first publication of "Autumnal Tints" in the *Atlantic Monthly* in 1862, was requested by Thoreau to illustrate his discussion of the leaf on pages 166-168.

NATURAL HISTORY OF MASSACHUSETTS [1]

Books of natural history make the most cheerful winter reading. I read in Audubon with a thrill of delight, when the snow covers the ground, of the magnolia, and the Florida keys, and their warm sea-breezes; of the fence-rail, and the cotton-tree, and the migrations of the rice-bird; of the breaking up of winter in Labrador, and the melting of the snow on the forks of the Missouri; and owe an accession of health to these reminiscences of luxuriant nature.

Within the circuit of this plodding life,
There enter moments of an azure hue,
Untarnished fair as is the violet
Or anemone, when the spring strews them
By some meandering rivulet, which make
The best philosophy untrue that aims
But to console man for his grievances.
I have remembered, when the winter came,
High in my chamber in the frosty nights,
When in the still light of the cheerful moon,
On every twig and rail and jutting spout,
The icy spears were adding to their length
Against the arrows of the coming sun,
How in the shimmering noon of summer past
Some unrecorded beam slanted across
The upland pastures where the Johnswort grew;
Or heard, amid the verdure of my mind,

[1] *Reports — on the Fishes, Reptiles, and Birds; the Herbaceous Plants and Quadrupeds; the Insects Injurious to Vegetation; and the Invertebrate Animals of Massachusetts.* Published agreeably to an Order of the Legislature, by the Commissioners on the Zoölogical and Botanical Survey of the State.

The bee's long smothered hum, on the blue flag
Loitering amidst the mead; or busy rill,
Which now through all its course stands still and dumb,
Its own memorial, — purling at its play
Along the slopes, and through the meadows next,
Until its youthful sound was hushed at last
In the staid current of the lowland stream;
Or seen the furrows shine but late upturned,
And where the fieldfare followed in the rear,
When all the fields around lay bound and hoar
Beneath a thick integument of snow.
So by God's cheap economy made rich
To go upon my winter's task again.

I am singularly refreshed in winter when I hear of service-berries, poke-weed, juniper. Is not heaven made up of these cheap summer glories? There is a singular health in those words, Labrador and East Main, which no desponding creed recognizes. How much more than Federal are these States! If there were no other vicissitudes than the seasons, our interest would never tire. Much more is adoing than Congress wots of. What journal do the persimmon and the buckeye keep, and the sharp-shinned hawk? What is transpiring from summer to winter in the Carolinas, and the Great Pine Forest, and the Valley of the Mohawk? The merely political aspect of the land is never very cheering; men are degraded when considered as the members of a political organization. On this side all lands present only the symptoms of decay. I see but Bunker Hill and Sing-Sing, the District of Columbia and Sullivan's Island, with a few avenues connecting them. But paltry are they all beside one blast of the east or the south wind which blows over them.

In society you will not find health, but in nature. Unless our feet at least stood in the midst of nature, all our faces would be pale and livid. Society is always diseased, and the best is the most so. There is no scent in it so wholesome as that of the pines, nor any fragrance so penetrating and restorative as the life-everlasting in high pastures. I would keep some book of natural history always by me as a sort of elixir, the reading of which should restore the tone of the system. To the sick, indeed, nature is sick, but to the well, a fountain of health. To him who contemplates a trait of natural beauty no harm nor disappointment can come. The doctrines of despair, of spiritual or political tyranny or servitude, were never taught by such as shared the serenity of nature. Surely good courage will not flag here on the Atlantic border, as long as we are flanked by the Fur Countries. There is enough in that sound to cheer one under any circumstances. The spruce, the hemlock, and the pine will not countenance despair. Methinks some creeds in vestries and churches do forget the hunter wrapped in furs by the Great Slave Lake, and that the Esquimaux sledges are drawn by dogs, and in the twilight of the northern night the hunter does not give over to follow the seal and walrus on the ice. They are of sick and diseased imaginations who would toll the world's knell so soon. Cannot these sedentary sects do better than prepare the shrouds and write the epitaphs of those other busy living men? The practical faith of all men belies the preacher's consolation. What is any man's discourse to me, if I am not sensible of something in it as steady and cheery as the creak of crickets? In it the woods must be relieved against the sky. Men tire me when I am not constantly greeted and refreshed as by the flux of sparkling streams. Surely joy is the condition of life. Think of the young fry that leap in ponds, the myriads of insects ushered into being on a summer evening, the incessant note of the hyla with which the woods ring in the spring, the nonchalance of the butterfly carrying accident and change painted in a thousand hues upon its wings, or the brook minnow stoutly stemming the current, the lustre of whose scales, worn bright by the attrition, is reflected upon the bank!

We fancy that this din of religion, literature, and philosophy, which is heard in pulpits, lyceums, and parlors, vibrates through the universe, and is as catholic a sound as the creaking of the earth's axle; but if a man sleep soundly, he will forget it all between sunset and dawn. It is the three-inch swing of a pendulum in a cupboard, which the great pulse of nature vibrates by and through each instant. When we lift our eyelids and open our ears, it disappears with smoke and rattle like the cars on a railroad. When I detect a beauty in any of the recesses of nature, I am reminded, by the serene and retired spirit in which it requires to be contemplated, of the inexpressible privacy of a life, — how silent and unambitious it is. The beauty there is in mosses must be considered from the holiest, quietest nook. What an admirable training is science for the more active warfare of life! Indeed, the unchallenged bravery which these studies imply, is far more impressive than the trumpeted valor of the warrior. I am pleased to learn that Thales was up and stirring by night not unfrequently, as his astronomical discoveries prove. Linnæus, setting out for Lapland, surveys his "comb" and "spare shirt," "leathern breeches" and "gauze cap to keep off gnats," with as much complacency as Bonaparte a park of artillery for the Russian campaign. The quiet bravery of the man is admirable. His eye is to take in fish, flower, and bird, quadruped and biped. Science is always brave; for to know is to know good; doubt and danger quail before her eye. What the coward overlooks in his hurry, she calmly scrutinizes, breaking ground like a pioneer for the array of arts that follow in her train. But cowardice is unscientific; for there cannot be a science of ignorance. There may be a science of bravery, for that advances; but a retreat is rarely well conducted; if it is, then is it an orderly advance in the face of circumstances.

But to draw a little nearer to our promised topics. Entomology extends the limits of being in a new direction, so that I walk in nature with a sense of greater space and freedom. It suggests besides, that the universe is not rough-hewn, but perfect in its details. Nature will bear the closest inspection; she invites us to lay our eye level with the smallest leaf, and take an insect view of its plain. She has no interstices; every part is full of life. I explore, too, with pleasure, the sources of the myriad sounds which crowd the summer noon, and which seem the very grain and stuff of which eternity is made. Who does not remember the shrill roll-call of the harvest-fly? There were ears for these sounds in Greece long ago, as Anacreon's ode will show.

"We pronounce thee happy, Cicada,
For on the tops of the trees,
Drinking a little dew,
Like any king thou singest,
For thine are they all,
Whatever thou seest in the fields,
And whatever the woods bear.
Thou art the friend of the husbandmen,
In no respect injuring any one;
And thou art honored among men,
Sweet prophet of summer.
The Muses love thee,
And Phœbus himself loves thee,
And has given thee a shrill song;
Age does not wrack thee,
Thou skillful, earthborn, song-loving,
Unsuffering, bloodless one;
Almost thou art like the gods."

In the autumn days, the creaking of crickets is heard at noon over all the land, and as in summer they are heard chiefly at nightfall, so then by their incessant chirp they usher in the evening of the year. Nor can all the vanities that vex the world alter one whit the measure that night has chosen. Every pulse-beat is in exact time with the cricket's chant and the tickings of the death-watch in the wall. Alternate with these if you can.

About two hundred and eighty birds either reside permanently in the State, or spend the summer only, or make us a passing visit. Those which spend the winter with us have obtained our warmest sympathy. The nut-hatch and chickadee flitting in company through the dells of the wood, the one harshly scolding at the intruder, the other with a faint lisping note enticing him on; the jay screaming in the orchard; the crow cawing in unison with the storm; the partridge, like a russet link extended over from autumn to spring, preserving unbroken the chain of summers; the hawk with warrior-like firmness abiding the blasts of winter; the robin [1] and lark lurking by warm springs in the woods; the familiar snowbird culling a few seeds in the garden or a few crumbs in the yard; and occasionally the shrike, with heedless and unfrozen melody bringing back summer again: —

His steady sails he never furls
At any time o' year,
And perching now on Winter's curls,
He whistles in his ear.

As the spring advances, and the ice is melting in the river, our earliest and straggling visitors make their appearance. Again does the old Teian poet sing as well for New England as for Greece, in the

RETURN OF SPRING

Behold, how, Spring appearing,
The Graces send forth roses;
Behold, how the wave of the sea
Is made smooth by the calm;
Behold, how the duck dives;
Behold, how the crane travels;

[1] A white robin and a white quail have occasionally been seen. It is mentioned in Audubon as remarkable that the nest of a robin should be found on the ground; but this bird seems to be less particular than most in the choice of a building-spot. I have seen its nest placed under the thatched roof of a deserted barn, and in one instance, where the adjacent country was nearly destitute of trees, together with two of the phœbe, upon the end of a board in the loft of a sawmill, but a few feet from the saw, which vibrated several inches with the motion of the machinery.

And Titan shines constantly bright.
The shadows of the clouds are moving;
The works of man shine;
The earth puts forth fruits;
The fruit of the olive puts forth.
The cup of Bacchus is crowned,
Along the leaves, along the branches,
The fruit, bending them down, flourishes.

The ducks alight at this season in the still water, in company with the gulls, which do not fail to improve an east wind to visit our meadows, and swim about by twos and threes, pluming themselves, and diving to peck at the root of the lily, and the cranberries which the frost has not loosened. The first flock of geese is seen beating to north, in long harrows and waving lines; the jingle of the song sparrow salutes us from the shrubs and fences; the plaintive note of the lark comes clear and sweet from the meadow; and the bluebird, like an azure ray, glances past us in our walk. The fish hawk, too, is occasionally seen at this season sailing majestically over the water, and he who has once observed it will not soon forget the majesty of its flight. It sails the air like a ship of the line, worthy to struggle with the elements, falling back from time to time like a ship on its beam ends, and holding its talons up as if ready for the arrows, in the attitude of the national bird. It is a great presence, as of the master of river and forest. Its eye would not quail before the owner of the soil, but make him feel like an intruder on its domains. And then its retreat, sailing so steadily away, is a kind of advance. I have by me one of a pair of ospreys, which have for some years fished in this vicinity, shot by a neighboring pond, measuring more than two feet in length, and six in the stretch of its wings. Nuttall mentions that "the ancients, particularly Aristotle, pretended that the ospreys taught their young to gaze at the sun, and those who were unable to do so were destroyed. Linnæus even believed, on ancient authority, that one of the feet of this bird had all the toes divided, while the other was partly webbed, so that it could swim with one foot, and grasp a fish with the other." But that educated eye is now dim, and those talons are nerveless. Its shrill scream seems yet to linger in its throat, and the roar of the sea in its wings. There is the tyranny of Jove in its claws, and his wrath in the erectile feathers of the head and neck. It reminds me of the Argonautic expedition, and would inspire the dullest to take flight over Parnassus.

The booming of the bittern, described by Goldsmith and Nuttall, is frequently heard in our fens, in the morning and evening, sounding like a pump, or the chopping of wood in a frosty morning in some distant farm-yard. The manner in which this sound is produced I have not seen anywhere described. On one occasion, the bird has been seen by one of my neighbors to thrust its bill into the water, and suck up as much as it could hold, then, raising its head, it pumped it out again with four or five heaves of the neck, throwing it two or three feet, and making the sound each time.

At length the summer's eternity is ushered in by the cackle of the flicker among the oaks on the hillside, and a new dynasty begins with calm security.

In May and June the woodland quire is in full tune, and, given the immense spaces of hollow air, and this curious human ear, one does not see how the void could be better filled.

> Each summer sound
> Is a summer round.

As the season advances, and those birds which make us but a passing visit depart, the woods become silent again, and but few feathers ruffle the drowsy air. But the solitary rambler may still find a response and expression for every mood in the depths of the wood.

> Sometimes I hear the veery's [1] clarion,
> Or brazen trump of the impatient jay,
> And in secluded woods the chickadee
> Doles out her scanty notes, which sing the praise
> Of heroes, and set forth the loveliness
> Of virtue evermore.

The phœbe still sings in harmony with the sultry weather by the brink of the pond, nor are the desultory hours of noon in the midst of the village without their minstrel.

> Upon the lofty elm-tree sprays
> The vireo rings the changes sweet,
> During the trivial summer days,
> Striving to lift our thoughts above the street.

With the autumn begins in some measure a new spring. The plover is heard whistling high in the air over the dry pastures, the finches flit from tree to tree,

[1] This bird, which is so well described by Nuttall, but is apparently unknown by the author of the Report, is one of the most common in the woods in this vicinity, and in Cambridge I have heard the college yard ring with its trill. The boys call it "yorrick," from the sound of its querulous and chiding note, as it flits near the traveler through the underwood. The cowbird's egg is occasionally found in its nest, as mentioned by Audubon.

the bobolinks and flickers fly in flocks, and the goldfinch rides on the earliest blast, like a winged hyla peeping amid the rustle of the leaves. The crows, too, begin now to congregate; you may stand and count them as they fly low and straggling over the landscape, singly or by twos and threes, at intervals of half a mile, until a hundred have passed.

I have seen it suggested somewhere that the crow was brought to this country by the white man; but I shall as soon believe that the white man planted these pines and hemlocks. He is no spaniel to follow our steps; but rather flits about the clearings like the dusky spirit of the Indian, reminding me oftener of Philip and Powhatan than of Winthrop and Smith. He is a relic of the dark ages. By just so slight, by just so lasting a tenure does superstition hold the world ever; there is the rook in England, and the crow in New England.

Thou dusky spirit of the wood,
Bird of an ancient brood,
Flitting thy lonely way,
A meteor in the summer's day,
From wood to wood, from hill to hill,
Low over forest, field, and rill,
What wouldst thou say?
Why shouldst thou haunt the day?
What makes thy melancholy float?
What bravery inspires thy throat,
And bears thee up above the clouds,
Over desponding human crowds,
Which far below
Lay thy haunts low?

The late walker or sailor, in the October evenings, may hear the murmurings of the snipe, circling over the meadows, the most spirit-like sound in nature; and still later in the autumn, when the frosts have tinged the leaves, a solitary loon pays a visit to our retired ponds, where he may lurk undisturbed till the season of moulting is passed, making the woods ring with his wild laughter. This bird, the Great Northern Diver, well deserves its name; for when pursued with a boat, it will dive, and swim like a fish under water, for sixty rods or more, as fast as a boat can be paddled, and its pursuer, if he would discover his game again, must put his ear to the surface to hear where it comes up. When it comes to the surface, it throws the water off with one shake of its wings, and calmly swims about until again disturbed.

These are the sights and sounds which reach our senses oftenest during the year. But sometimes one hears a quite new note, which has for background other Carolinas and Mexicos than the books describe, and learns that his ornithology has done him no service.

It appears from the Report that there are about forty quadrupeds belonging to the State, and among these one is glad to hear of a few bears, wolves, lynxes, and wildcats.

When our river overflows its banks in the spring, the wind from the meadows is laden with a strong scent of musk, and by its freshness advertises me of an unexplored wildness. Those backwoods are not far off then. I am affected by the sight of the cabins of the muskrat, made of mud and grass, and raised three or four feet along the river, as when I read of the barrows of Asia. The muskrat is the beaver of the settled States. Their number has even increased within a few years in this vicinity. Among the rivers which empty into the Merrimack, the Concord is known to the boatmen as a dead stream. The Indians are said to have called it Musketaquid, or Prairie River. Its current being much more sluggish and its water more muddy than the rest, it abounds more in fish and game of every kind. According to the History of the town, "The fur-trade was here once very important. As early as 1641, a company was formed in the colony, of which Major Willard of Concord was superintendent, and had the exclusive right to trade with the Indians in furs and other articles; and for this right they were obliged to pay into the public treasury one twentieth of all the furs they obtained." There are trappers in our midst still, as well as on the streams of the far West, who night and morning go the round of their traps, without fear of the Indian. One of these takes from one hundred and fifty to two hundred muskrats in a year, and even thirty-six have been shot by one man in a day. Their fur, which is not nearly as valuable as formerly, is in good condition in the winter and spring only; and upon the breaking up of the ice, when they are driven out of their holes by the water, the greatest number is shot from boats, either swimming or resting on their stools, or slight supports of grass and reeds, by the side of the stream. Though they exhibit considerable cunning at other times, they are easily taken in a trap, which has only to be placed in their holes, or wherever they frequent, without any bait being used, though it is sometimes rubbed with their musk. In the winter the hunter cuts holes in the ice, and shoots them when they come to the surface. Their burrows are usually in the high banks of the river, with the entrance under water, and rising within to above the level of high water. Sometimes their nests, composed of dried meadow-grass and flags, may be discovered where the bank is low and spongy, by the yielding of the ground under the feet. They have from three to seven or eight young in the spring.

Frequently, in the morning or evening, a long ripple is seen in the still water, where a muskrat is crossing the stream, with only its nose above the surface, and sometimes a green bough in its mouth to build its house with. When it finds itself observed, it will dive and swim five or six rods under water, and at length conceal itself in its hole, or the weeds. It will remain under water for ten minutes at a time, and on one occasion has been seen, when undisturbed, to form an air-bubble under the ice, which contracted and expanded as it breathed at leisure. When it suspects danger on shore, it will stand erect like a squirrel, and survey its neighborhood for several minutes, without moving.

In the fall, if a meadow intervene between their burrows and the stream, they erect cabins of mud and grass, three or four feet high, near its edge. These are not their breeding-places, though young are sometimes found in them in late freshets, but rather their hunting-lodges, to which they resort in the winter with their food, and for shelter. Their food consists chiefly of flags and fresh-water mussels, the shells of the latter being left in large quantities around their lodges in the spring.

The Penobscot Indian wears the entire skin of a muskrat, with the legs and tail dangling, and the head caught under his girdle, for a pouch, into which he puts his fishing-tackle, and essences to scent his traps with.

The bear, wolf, lynx, wildcat, deer, beaver, and marten have disappeared; the otter is rarely if ever seen here at present; and the mink is less common than formerly.

Perhaps of all our untamed quadrupeds, the fox has obtained the widest and most familiar reputation, from the time of Pilpay and Æsop to the present day. His recent tracks still give variety to a winter's walk. I tread in the steps of the fox that has gone before me by some hours, or which perhaps I have started, with such a tip-toe of expectation as if I were on the trail of the Spirit itself which resides in the wood, and expected soon to catch it in its lair. I am curious to know what has determined its graceful curvatures, and how surely they were coincident with the fluctuations of some mind. I know which way a mind wended, what horizon it faced, by the setting of these tracks, and whether it moved slowly or rapidly, by their greater or less intervals and distinctness; for the swiftest step leaves yet a lasting trace. Sometimes you will see the trails of many together, and where they have gamboled and gone through a hundred evolutions, which testify to a singular listlessness and leisure in nature.

When I see a fox run across the pond on the snow, with the carelessness of freedom, or at intervals trace his course in the sunshine along the ridge of a hill, I give up to him sun and earth as to their true proprietor. He does not go in the sun, but it seems to follow him, and there is a visible sympathy between him and it.

Sometimes, when the snow lies light and but five or six inches deep, you may give chase and come up with one on foot. In such a case he will show a remarkable presence of mind, choosing only the safest direction,though he may lose ground by it. Notwithstanding his fright, he will take no step which is not beautiful. His pace is a sort of leopard canter, as if he were in no wise impeded by the snow, but were husbanding his strength all the while. When the ground is uneven, the course is a series of graceful curves, conforming to the shape of the surface. He runs as though there were not a bone in his back. Occasionally dropping his muzzle to the ground for a rod or two, and then tossing his head aloft, when satisfied of his course. When he comes to a declivity, he will put his fore feet together, and slide swiftly down it, shoving the snow before him. He treads so softly that you would hardly hear it from any nearness, and yet with such expression that it would not be quite inaudible at any distance.

Of fishes, seventy-five genera and one hundred and seven species are described in the Report. The fisherman will be startled to learn that there are but about a dozen kinds in the ponds and streams of any inland town; and almost nothing is known of their habits. Only their names and residence make one love fishes. I would know even the number of their fin-rays, and how many scales compose the lateral line. I am the wiser in respect to all knowledges, and the better qualified for all fortunes, for knowing that there is a minnow in the brook. Methinks I have need even of his sympathy, and to be his fellow in a degree.

I have experienced such simple delight in the trivial matters of fishing and sporting, formerly, as might have inspired the muse of Homer or Shakespeare; and now, when I turn the pages and ponder the plates of the Angler's Souvenir, I am fain to exclaim, —

> "Can such things be,
> And overcome us like a summer's cloud?"

Next to nature, it seems as if man's actions were the most natural, they so gently accord with her. The small seines of flax stretched across the shallow and transparent parts of our river are no more intrusion than the cobweb in the sun. I stay my boat in mid-current, and look down in the sunny water to see the civil meshes of his nets, and wonder how the blustering people of the town could have done this elvish work. The twine looks like a new river-weed, and is to the river as a beautiful memento of man's presence in nature, discovered as silently and delicately as a footprint in the sand.

When the ice is covered with snow, I do not suspect the wealth under my feet; that there is as good as a mine under me wherever I go. How many pickerel are poised on easy fin fathoms below the loaded wain! The revolution of the seasons must be a curious phenomenon to them. At length the sun and wind brush aside their curtain, and they see the heavens again.

Early in the spring, after the ice has melted, is the time for spearing fish. Suddenly the wind shifts from northeast and east to west and south, and every icicle, which has tinkled on the meadow grass so long, trickles down its stem, and seeks its level unerringly with a million comrades. The steam curls up from every roof and fence.

I see the civil sun drying earth's tears,
Her tears of joy, which only faster flow.

In the brooks is heard the slight grating sound of small cakes of ice, floating with various speed, full of content and promise, and where the water gurgles under a natural bridge, you may hear these hasty rafts hold conversation in an undertone. Every rill is a channel for the juices of the meadow. In the ponds the ice cracks with a merry and inspiriting din, and down the larger streams is whirled grating hoarsely, and crashing its way along, which was so lately a highway for the woodman's team and the fox, sometimes with the tracks of the skaters still fresh upon it, and the holes cut for pickerel. Town committees anxiously inspect the bridges and causeways, as if by mere eye-force to intercede with the ice and save the treasury.

The river swelleth more and more,
Like some sweet influence stealing o'er
The passive town; and for a while
Each tussock makes a tiny isle,
Where, on some friendly Ararat,
Resteth the weary water-rat.

No ripple shows Musketaquid,
Her very current e'en is hid,
As deepest souls do calmest rest
When thoughts are swelling in the breast,
And she that in the summer's drought
Doth make a rippling and a rout,
Sleeps from Nahshawtuck to the Cliff,
Unruffled by a single skiff.
But by a thousand distant hills
The louder roar a thousand rills,

And many a spring which now is dumb,
And many a stream with smothered hum,
Doth swifter well and faster glide,
Though buried deep beneath the tide.
Our village shows a rural Venice,
Its broad lagoons where yonder fen is;
As lovely as the Bay of Naples
Yon placid cove amid the maples;
And in my neighbor's field of corn
I recognize the Golden Horn.

Here Nature taught from year to year,
When only red men came to hear, —
Methinks 't was in this school of art
Venice and Naples learned their part;
But still their mistress, to my mind,
Her young disciples leaves behind.

The fisherman now repairs and launches his boat. The best time for spearing is at this season, before the weeds have begun to grow, and while the fishes lie in the shallow water, for in summer they prefer the cool depths, and in the autumn they are still more or less concealed by the grass. The first requisite is fuel for your crate; and for this purpose the roots of the pitch pine are commonly used, found under decayed stumps, where the trees have been felled eight or ten years.

With a crate, or jack, made of iron hoops, to contain your fire, and attached to the bow of your boat about three feet from the water, a fish-spear with seven tines and fourteen feet long, a large basket or barrow to carry your fuel and bring back your fish, and a thick outer garment, you are equipped for a cruise. It should be a warm and still evening; and then, with a fire crackling merrily at the prow, you may launch forth like a cucullo into the night. The dullest soul cannot go upon such an expedition without some of the spirit of adventure; as if he had stolen the boat of Charon and gone down the Styx on a midnight expedition into the realms of Pluto. And much speculation does this wandering star afford to the musing night-walker, leading him on and on, jack-o'-lantern-like, over the meadows; or, if he is wiser, he amuses himself with imagining what of human life, far in the silent night, is flitting moth-like round its candle. The silent navigator shoves his craft gently over the water, with a smothered pride and sense of benefaction, as if he were the phosphor, or light-bringer, to these dusky realms, or some sister moon, blessing the spaces with her light. The waters, for a rod or two on either hand and several feet in depth, are lit up with more than noonday distinctness, and he enjoys the opportunity which so many have desired, for the roofs of a city are indeed raised, and he surveys the midnight economy of the fishes. There they lie in every variety of posture; some on their backs, with their white bellies uppermost, some suspended in mid-water, some sculling gently along with a dreamy motion of the fins, and others quite active and wide awake, — a scene not unlike what the human city would present. Occasionally he will encounter a turtle selecting the choicest morsels, or a muskrat resting on a tussock. He may exercise his dexterity, if he sees fit, on the more distant and active fish, or fork the nearer into his boat, as potatoes out of a pot, or even take the sound sleepers with his hands. But these last accomplishments he will soon learn to dispense with, distinguishing the real object of his pursuit, and find compensation in the beauty and never-ending novelty of his position. The pines growing down to the water's edge will show newly as in the glare of a conflagration; and as he floats under the willows with his light, the song sparrow will often wake on her perch, and sing that strain at midnight which she had meditated for the morning. And when he has done, he may have to steer his way home through the dark by the north star, and he will feel himself some degrees nearer to it for having lost his way on the earth.

The fishes commonly taken in this way are pickerel, suckers, perch, eels, pouts, breams, and shiners, — from thirty to sixty weight in a night. Some are hard to be recognized in the unnatural light, especially the perch, which, his dark bands being exaggerated, acquires a ferocious aspect. The number of these transverse bands, which the Report states to be seven, is, however, very variable, for in some of our ponds they have nine and ten even.

It appears that we have eight kinds of tortoises, twelve snakes, — but one of which is venomous, — nine frogs and toads, nine salamanders, and one lizard, for our neighbors.

I am particularly attracted by the motions of the serpent tribe. They make our hands and feet, the wings of the bird, and the fins of the fish seem very superfluous, as if Nature had only indulged her fancy in making them. The black snake will dart into a bush when pursued, and circle round and round with an easy and graceful motion, amid the thin and bare twigs, five or six feet from the ground, as a bird flits from bough to bough, or hang in festoons between the forks. Elasticity and flexibleness in the simpler forms of animal life are equivalent to a complex system of limbs in the higher; and we have only to be as wise and wily as the serpent, to perform as difficult feats without the vulgar assistance of hands and feet.

In May, the snapping turtle (*Emysaurus serpentina*) is frequently taken on the meadows and in the river. The fisherman, taking sight over the calm surface, discovers its snout projecting above the water, at the distance of many rods, and easily secures his prey through its unwillingness to disturb the water by swimming hastily away, for, gradually drawing its head under, it remains resting on some limb or clump of grass. Its eggs, which are buried at a distance from the water, in some soft place, as a pigeon-bed, are frequently devoured by the skunk. It will catch fish by daylight, as a toad catches flies, and is said to emit a transparent fluid from its mouth to attract them.

Nature has taken more care than the fondest parent for the education and refinement of her children. Consider the silent influence which flowers exert, no less upon the ditcher in the meadow than the lady in the bower. When I walk in the woods, I am reminded that a wise purveyor has been there before me; my most delicate experience is typified there. I am struck with the pleasing friendships and unanimities of nature, as when the lichen on the trees takes the form of their leaves. In the most stupendous scenes you will see delicate and fragile features, as slight wreaths of vapor, dew-lines, feathery sprays, which suggest a high refinement, a noble blood and breeding, as it were. It is not hard to account for elves and fairies; they represent this light grace, this ethereal gentility. Bring a spray from the wood, or a crystal from the brook, and place it on your mantel, and your household ornaments will seem plebeian beside its nobler fashion and bearing. It will wave superior there, as if used to a more refined and polished circle. It has a salute and a response to all your enthusiasm and heroism.

In the winter, I stop short in the path to admire how the trees grow up without forethought, regardless of the time and circumstances. They do not wait as man does, but now is the golden age of the sapling. Earth, air, sun, and rain are occasion enough; they were no better in primeval centuries. The "winter of *their* discontent" never comes. Witness the buds of the native poplar standing gayly out to the frost on the sides of its bare switches. They express a naked confidence. With cheerful heart one could be a sojourner in the wilderness, if he were sure to find there the catkins of the willow or the alder. When I read of them in the accounts of northern adventurers, by Baffin's Bay or Mackenzie's River, I see how even there, too, I could dwell. They are our little vegetable redeemers. Methinks our virtue will hold out till they come again. They are worthy to have had a greater than Minerva or Ceres for their inventor. Who was the benignant goddess that bestowed them on mankind?

Nature is mythical and mystical always, and works with the license and extravagance of genius. She has her luxurious and florid style as well as art. Having a pilgrim's cup to make, she gives to the whole — stem, bowl, handle, and nose — some fantastic shape, as if it were to be the car of some fabulous marine deity, a Nereus or Triton.

In the winter, the botanist need not confine himself to his books and herbarium, and give over his outdoor pursuits, but may study a new department of vegetable physiology, what may be called crystalline botany, then. The winter of 1837 was unusually favorable for this. In December of that year, the Genius of vegetation seemed to hover by night over its summer haunts with unusual persistency. Such a hoar-frost as is very uncommon here or anywhere, and whose full effects can never be witnessed after sunrise, occurred several times. As I went forth early on a still and frosty morning, the trees looked like airy creatures of darkness caught napping; on this side huddled together, with their gray hairs streaming, in a secluded valley which the sun had not penetrated; on that, hurrying off in Indian file along some watercourse, while the shrubs and grasses, like elves and fairies of the night, sought to hide their diminished heads in the snow. The river, viewed from the high bank, appeared of a yellowish-green color, though all the landscape was white. Every tree, shrub, and spire of grass, that could raise its head above the snow, was covered with a dense ice-foliage, answering, as it were, leaf for leaf to its summer dress. Even the fences had put forth leaves in the night. The centre, diverging, and more minute fibres were perfectly distinct, and the edges regularly indented. These leaves were on the side of the twig or stubble opposite to the sun, meeting it for the most part at right angles, and there were others standing out at all possible angles upon these and upon one another, with no twig or stubble supporting them. When the first rays of the sun slanted over the scene, the grasses seemed hung with innumerable jewels, which jingled merrily as they were brushed by the foot of the traveler, and reflected all the hues of the rainbow, as he moved from side to side. It struck me that these ghost leaves, and the green ones whose forms they assume, were the creatures of but one law; that in obedience to the same law the vegetable juices swell gradually into the perfect leaf, on the one hand, and the crystalline particles troop to their standard in the same order, on the other. As if the material were indifferent, but the law one and invariable, and every plant in the spring but pushed up into and filled a permanent and eternal mould, which, summer and winter forever, is waiting to be filled.

This foliate structure is common to the coral and the plumage of birds, and to how large a part of animate and inanimate nature. The same independence of law on matter is observable in many other instances, as in the natural rhymes, when some animal form, color, or odor has its counterpart in some vegetable. As, indeed, all rhymes imply an eternal melody, independent of any particular sense.

As confirmation of the fact that vegetation is but a kind of crystallization, every one may observe how, upon the edge of the melting frost on the window, the needle-shaped particles are bundled together so as to resemble fields waving with grain, or shocks rising here and there from the stubble; on one side the vegetation of the torrid zone, high-towering palms and wide-spread banyans, such as are seen in pictures of oriental scenery; on the other, arctic pines stiff frozen, with downcast branches.

Vegetation has been made the type of all growth; but as in crystals the law is more obvious, their material being more simple, and for the most part more transient and fleeting, would it not be as philosophical as convenient to consider all growth, all filling up within the limits of nature, but a crystallization more or less rapid?

On this occasion, in the side of the high bank of the river, wherever the water or other cause had formed a cavity, its throat and outer edge, like the entrance to a citadel, bristled with a glistening ice-armor. In one place you might see minute ostrich-feathers, which seemed the waving plumes of the warriors filing into the fortress; in another, the glancing, fan-shaped banners of the Lilliputian host; and in another, the needle-shaped particles collected into bundles, resembling the plumes of the pine, might pass for a phalanx of spears. From the under side of the ice in the brooks, where there was a thicker ice below, depended a mass of crystallization, four or five inches deep, in the form of prisms, with their lower ends open, which, when the ice was laid on its smooth side, resembled the roofs and steeples of a Gothic city, or the vessels of a crowded haven under a press of canvas. The very mud in the road, where the ice had melted, was crystallized with deep rectilinear fissures, and the crystalline masses in the sides of the ruts resembled exactly asbestos in the disposition of their needles. Around the roots of the stubble and flower-stalks, the frost was gathered into the form of irregular conical shells, or fairy rings. In some places the ice-crystals were lying upon granite rocks, directly over crystals of quartz, the frostwork of a longer night, crystals of a longer period, but, to some eye unprejudiced by the short term of human life, melting as fast as the former.

In the Report on the Invertebrate Animals, this singular fact is recorded, which teaches us to put a new value on time and space: "The distribution of the marine shells is well worthy of notice as a geological fact. Cape Cod, the right arm of the Commonwealth, reaches out into the ocean, some fifty or sixty miles. It is nowhere many miles wide; but this narrow point of land has hitherto proved a barrier to the migrations of many species of Mollusca. Several genera and numerous species, which are separated by the intervention of only a few miles of land, are effectually prevented from mingling by the Cape, and do not pass from one side to the other. . . . Of the one hundred and ninety-seven marine species, eighty-three do not pass to the south shore, and fifty are not found on the north shore of the Cape."

That common mussel, the *Unio complanatus*, or more properly *fluviatilis*, left in the spring by the muskrat upon rocks and stumps, appears to have been an important article of food with the Indians. In one place, where they are said to have feasted, they are found in large quantities, at an elevation of thirty feet above the river, filling the soil to the depth of a foot, and mingled with ashes and Indian remains.

The works we have placed at the head of our chapter, with as much license as the preacher selects his text, are such as imply more labor than enthusiasm. The State wanted complete catalogues of its natural riches, with such additional facts merely as would be directly useful.

The reports on Fishes, Reptiles, Insects, and Invertebrate Animals, however, indicate labor and research, and have a value independent of the object of the legislature.

Those on Herbaceous Plants and Birds cannot be of much value, as long as Bigelow and Nuttall are accessible. They serve but to indicate, with more or less exactness, what species are found in the State. We detect several errors ourselves, and a more practiced eye would no doubt expand the list.

The Quadrupeds deserved a more final and instructive report than they have obtained.

These volumes deal much in measurements and minute descriptions, not interesting to the general reader, with only here and there a colored sentence to allure him, like those plants growing in dark forests, which bear only leaves without blossoms. But the ground was comparatively unbroken, and we will not complain of the pioneer, if he raises no flowers with his first crop. Let us not underrate the value of a fact; it will one day flower in a truth. It is astonishing how few facts of importance are added in a century to the natural history of any animal. The natural history of man himself is still being gradually written. Men are knowing enough after their fashion. Every countryman and dairy-maid knows that the coats of the fourth stomach of the calf will curdle milk, and what particular mushroom is a safe and nutritious diet. You cannot go into any field or wood, but it will seem as if every stone had been turned, and the bark on every tree ripped up. But, after all, it is much easier to discover than to see when the cover is off. It has been well said that "the attitude of inspection is prone." Wisdom does not inspect, but behold. We must look a long time before we can see. Slow are the beginnings of philosophy. He has something demoniacal in him, who can discern a law or couple two facts. We can imagine a time when "Water runs down hill" may have been taught in the schools. The true man of science will know nature better by his finer organization; he will smell, taste, see, hear, feel, better than other men. His will be a deeper and finer experience. We do not learn by inference and deduction and the application of mathematics to philosophy, but by direct intercourse and sympathy. It is with science as with ethics, — we cannot know truth by contrivance and method; the Baconian is as false as any other, and with all the helps of machinery and the arts, the most scientific will still be the healthiest and friendliest man, and possess a more perfect Indian wisdom.

A WALK TO WACHUSETT

CONCORD, July 19, 1842.

The needles of the pine
All to the west incline.

SUMMER and winter our eyes had rested on the dim outline of the mountains in our horizon, to which distance and indistinctness lent a grandeur not their own, so that they served equally to interpret all the allusions of poets and travelers; whether with Homer, on a spring morning, we sat down on the many-peaked Olympus, or with Virgil and his compeers roamed the Etrurian and Thessalian hills, or with Humboldt measured the more modern Andes and Teneriffe. Thus we spoke our mind to them, standing on the Concord cliffs: —

With frontier strength ye stand your ground,
With grand content ye circle round,
Tumultuous silence for all sound,
Ye distant nursery of rills,
Monadnock, and the Peterboro' hills;
Like some vast fleet,
Sailing through rain and sleet,
Through winter's cold and summer's heat;
Still holding on, upon your high emprise,
Until ye find a shore amid the skies;
Not skulking close to land,
With cargo contraband,
For they who sent a venture out by ye
Have set the sun to see
Their honesty.
Ships of the line, each one,
Ye to the westward run,

Always before the gale,
Under a press of sail,
With weight of metal all untold.
I seem to feel ye, in my firm seat here,
Immeasurable depth of hold,
And breadth of beam, and length of running gear.

Methinks ye take luxurious pleasure
In your novel western leisure;
So cool your brows, and freshly blue,
As Time had nought for ye to do;
For ye lie at your length,
An unappropriated strength,
Unhewn primeval timber,
For knees so stiff, for masts so limber;
The stock of which new earths are made
One day to be our western trade,
Fit for the stanchions of a world
Which through the seas of space is hurled.

While we enjoy a lingering ray,
Ye still o'ertop the western day,
Reposing yonder, on God's croft,
Like solid stacks of hay.
Edged with silver, and with gold,
The clouds hang o'er in damask fold,
And with such depth of amber light
The west is dight,
Where still a few rays slant,
That even heaven seems extravagant.
On the earth's edge mountains and trees
Stand as they were on air graven,
Or as the vessels in a haven
Await the morning breeze.
I fancy even
Through your defiles windeth the way to heaven;
And yonder still, in spite of history's page,
Linger the golden and the silver age;
Upon the laboring gale

The news of future centuries is brought,
And of new dynasties of thought,
From your remotest vale.

But special I remember thee,
Wachusett, who like me
Standest alone without society.
Thy far blue eye,
A remnant of the sky,
Seen through the clearing or the gorge
Or from the windows of the forge,
Doth leaven all it passes by.
Nothing is true,
But stands 'tween me and you,
Thou western pioneer,
Who know'st not shame nor fear
By venturous spirit driven,
Under the eaves of heaven.
And canst expand thee there,
And breathe enough of air?
Upholding heaven, holding down earth,
Thy pastime from thy birth,
Not steadied by the one, nor leaning on the other;
May I approve myself thy worthy brother!

At length, like Rasselas, and other inhabitants of happy valleys, we resolved to scale the blue wall which bounded the western horizon, though not without misgivings that thereafter no visible fairyland would exist for us. But we will not leap at once to our journey's end, though near, but imitate Homer, who conducts his reader over the plain, and along the resounding sea, though it be but to the tent of Achilles. In the spaces of thought are the reaches of land and water, where men go and come. The landscape lies far and fair within, and the deepest thinker is the farthest traveled.

At a cool and early hour on a pleasant morning in July, my companion and I passed rapidly through Acton and Stow, stopping to rest and refresh us on the bank of a small stream, a tributary of the Assabet, in the latter town. As we traversed the cool woods of Acton, with stout staves in our hands, we were cheered by the song of the red-eye, the thrushes, the phœbe, and the cuckoo; and as we passed through the open country, we inhaled the fresh scent of every field, and all nature lay passive, to be viewed and traveled. Every rail, every farmhouse, seen dimly in the twilight, every tinkling sound told of peace and purity, and we moved happily along the dank roads, enjoying not such privacy as the day leaves when it withdraws, but such as it has not profaned. It was solitude with light; which is better than darkness. But anon, the sound of the mower's rifle was heard in the fields, and this, too, mingled with the lowing of kine.

This part of our route lay through the country of hops, which plant perhaps supplies the want of the vine in American scenery, and may remind the traveler of Italy and the South of France, whether he traverses the country when the hop-fields, as then, present solid and regular masses of verdure, hanging in graceful festoons from pole to pole, the cool coverts where lurk the gales which refresh the wayfarer; or in September, when the women and children, and the neighbors from far and near, are gathered to pick the hops into long troughs; or later still, when the poles stand piled in vast pyramids in the yards, or lie in heaps by the roadside.

The culture of the hop, with the processes of picking,

drying in the kiln, and packing for the market, as well as the uses to which it is applied, so analogous to the culture and uses of the grape, may afford a theme for future poets.

The mower in the adjacent meadow could not tell us the name of the brook on whose banks we had rested, or whether it had any, but his younger companion, perhaps his brother, knew that it was Great Brook. Though they stood very near together in the field, the things they knew were very far apart; nor did they suspect each other's reserved knowledge, till the stranger came by. In Bolton, while we rested on the rails of a cottage fence, the strains of music which issued from within, probably in compliment to us, sojourners, reminded us that thus far men were fed by the accustomed pleasures. So soon did we, wayfarers, begin to learn that man's life is rounded with the same few facts, the same simple relations everywhere, and it is vain to travel to find it new. The flowers grow more various ways than he. But coming soon to higher land, which afforded a prospect of the mountains, we thought we had not traveled in vain, if it were only to hear a truer and wilder pronunciation of their names from the lips of the inhabitants; not *Way*-tatic, *Way*-chusett, but *Wor*-tatic, *Wor*-chusett. It made us ashamed of our tame and civil pronunciation, and we looked upon them as born and bred farther west than we. Their tongues had a more generous accent than ours, as if breath was cheaper where they wagged. A countryman, who speaks but seldom, talks copiously, as it were, as his wife sets cream and cheese before you without stint. Before noon we had reached the highlands overlooking the valley of Lancaster (affording the first fair and open prospect into the west), and there, on the top of a hill, in the shade of some oaks, near to where a spring bubbled out from a leaden pipe, we rested during the heat of the day, reading Virgil and enjoying the scenery. It was such a place as one feels to be on the outside of the earth; for from it we could, in some measure, see the form and structure of the globe. There lay Wachusett, the object of our journey, lowering upon us with unchanged proportions, though with a less ethereal aspect than had greeted our morning gaze, while further north, in successive order, slumbered its sister mountains along the horizon.

We could get no further into the Æneid than

— atque altae moenia Romae,
— and the wall of high Rome,

before we were constrained to reflect by what myriad tests a work of genius has to be tried; that Virgil, away in Rome, two thousand years off, should have to unfold his meaning, the inspiration of Italian vales, to the pilgrim on New England hills. This life so raw and modern, that so civil and ancient; and yet we read Virgil mainly to be reminded of the identity of human nature in all ages, and, by the poet's own account, we are both the children of a late age, and live equally under the reign of Jupiter.

"He shook honey from the leaves, and removed fire,
And stayed the wine, everywhere flowing in rivers;
That experience, by meditating, might invent various arts
By degrees, and seek the blade of corn in furrows,
And strike out hidden fire from the veins of the flint."

The old world stands serenely behind the new, as

one mountain yonder towers behind another, more dim and distant. Rome imposes her story still upon this late generation. The very children in the school we had that morning passed had gone through her wars, and recited her alarms, ere they had heard of the wars of neighboring Lancaster. The roving eye still rests inevitably on her hills, and she still holds up the skirts of the sky on that side, and makes the past remote.

The lay of the land hereabouts is well worthy the attention of the traveler. The hill on which we were resting made part of an extensive range, running from southwest to northeast, across the country, and separating the waters of the Nashua from those of the Concord, whose banks we had left in the morning, and by bearing in mind this fact, we could easily determine whither each brook was bound that crossed our path. Parallel to this, and fifteen miles further west, beyond the deep and broad valley in which lie Groton, Shirley, Lancaster, and Boylston, runs the Wachusett range, in the same general direction. The descent into the valley on the Nashua side is by far the most sudden; and a couple of miles brought us to the southern branch of the Nashua, a shallow but rapid stream, flowing between high and gravelly banks. But we soon learned that these were no *gelidae valles* into which we had descended, and, missing the coolness of the morning air, feared it had become the sun's turn to try his power upon us.

> " The sultry sun had gained the middle sky,
> And not a tree, and not an herb was nigh,"

and with melancholy pleasure we echoed the melodious plaint of our fellow-traveler, Hassan, in the desert, —

"Sad was the hour, and luckless was the day,
When first from Schiraz' walls I bent my way."

The air lay lifeless between the hills, as in a seething caldron, with no leaf stirring, and instead of the fresh odor of grass and clover, with which we had before been regaled, the dry scent of every herb seemed merely medicinal. Yielding, therefore, to the heat, we strolled into the woods, and along the course of a rivulet, on whose banks we loitered, observing at our leisure the products of these new fields. He who traverses the woodland paths, at this season, will have occasion to remember the small, drooping, bell-like flowers and slender red stem of the dogsbane, and the coarser stem and berry of the poke, which are both common in remoter and wilder scenes; and if "the sun casts such a reflecting heat from the sweet-fern" as makes him faint, when he is climbing the bare hills, as they complained who first penetrated into these parts, the cool fragrance of the swamp-pink restores him again, when traversing the valleys between.

As we went on our way late in the afternoon, we refreshed ourselves by bathing our feet in every rill that crossed the road, and anon, as we were able to walk in the shadows of the hills, recovered our morning elasticity. Passing through Sterling, we reached the banks of the Stillwater, in the western part of the town, at evening, where is a small village collected. We fancied that there was already a certain western look about this place, a smell of pines and roar of water, recently confined by dams, belying its name, which were exceedingly grateful. When the first inroad has been made, a few acres leveled, and a few houses erected, the forest looks wilder than ever. Left to herself, nature is always more or less civilized, and delights in a certain refinement; but where the axe has encroached upon the edge of the forest, the dead and unsightly limbs of the pine, which she had concealed with green banks of verdure, are exposed to sight. This village had, as yet, no post-office, nor any settled name. In the small villages which we entered, the villagers gazed after us, with a complacent, almost compassionate look, as if we were just making our *début* in the world at a late hour. "Nevertheless," did they seem to say, "come and study us, and learn men and manners." So is each one's world but a clearing in the forest, so much open and inclosed ground. The landlord had not yet returned from the field with his men, and the cows had yet to be milked. But we remembered the inscription on the wall of the Swedish inn, "You will find at Trolhate excellent bread, meat, and wine, provided you bring them with you," and were contented. But I must confess it did somewhat disturb our pleasure, in this withdrawn spot, to have our own village newspaper handed us by our host, as if the greatest charm the country offered to the traveler was the facility of communication with the town. Let it recline on its own everlasting hills, and not be looking out from their summits for some petty Boston or New York in the horizon.

At intervals we heard the murmuring of water, and the slumberous breathing of crickets, throughout the night; and left the inn the next morning in the gray twilight, after it had been hallowed by the night air, and when only the innocent cows were stirring, with a kind of regret. It was only four miles to the base of the mountain, and the scenery was already more picturesque. Our road lay along the course of the Stillwater, which was brawling at the bottom of a deep ravine, filled with pines and rocks, tumbling fresh from the mountains, so soon, alas! to commence its career of usefulness. At first, a cloud hung between us and the summit, but it was soon blown away. As we gathered the raspberries, which grew abundantly by the roadside, we fancied that that action was consistent with a lofty prudence; as if the traveler who ascends into a mountainous region should fortify himself by eating of such light ambrosial fruits as grow there, and drinking of the springs which gush out from the mountain-sides, as he gradually inhales the subtler and purer atmosphere of those elevated places, thus propitiating the mountain gods by a sacrifice of their own fruits. The gross products of the plains and valleys are for such as dwell therein; but it seemed to us that the juices of this berry had relation to the thin air of the mountain-tops.

In due time we began to ascend the mountain, passing, first, through a grand sugar maple wood, which bore the marks of the auger, then a denser forest, which gradually became dwarfed, till there were no trees whatever. We at length pitched our tent on the summit. It is but nineteen hundred feet above the village of Princeton, and three thousand above the level of the sea; but by this slight elevation it is infinitely removed from the plain, and when we reached it we felt a sense of remoteness, as if we had traveled into distant regions, to Arabia

Petræa, or the farthest East. A robin upon a staff was the highest object in sight. Swallows were flying about us, and the chewink and cuckoo were heard near at hand. The summit consists of a few acres, destitute of trees, covered with bare rocks, interspersed with blueberry bushes, raspberries, gooseberries, strawberries, moss, and a fine, wiry grass. The common yellow lily and dwarf cornel grow abundantly in the crevices of the rocks. This clear space, which is gently rounded, is bounded a few feet lower by a thick shrubbery of oaks, with maples, aspens, beeches, cherries, and occasionally a mountain-ash intermingled, among which we found the bright blue berries of the Solomon's-seal, and the fruit of the pyrola. From the foundation of a wooden observatory, which was formerly erected on the highest point, forming a rude, hollow structure of stone, a dozen feet in diameter, and five or six in height, we could see Monadnock, in simple grandeur, in the northwest, rising nearly a thousand feet higher, still the "far blue mountain," though with an altered profile. The first day the weather was so hazy that it was in vain we endeavored to unravel the obscurity. It was like looking into the sky again, and the patches of forest here and there seemed to flit like clouds over a lower heaven. As to voyagers of an aerial Polynesia, the earth seemed like a larger island in the ether; on every side, even as low as we, the sky shutting down, like an unfathomable deep, around it, a blue Pacific island, where who knows what islanders inhabit? and as we sail near its shores we see the waving of trees and hear the lowing of kine.

We read Virgil and Wordsworth in our tent, with

new pleasure there, while waiting for a clearer atmosphere, nor did the weather prevent our appreciating the simple truth and beauty of Peter Bell: —

"And he had lain beside his asses,
On lofty Cheviot Hills:

"And he had trudged through Yorkshire dales,
Among the rocks and winding *scars;*
Where deep and low the hamlets lie
Beneath their little patch of sky
And little lot of stars."

Who knows but this hill may one day be a Helvellyn, or even a Parnassus, and the Muses haunt here, and other Homers frequent the neighboring plains?

Not unconcerned Wachusett rears his head
Above the field, so late from nature won,
With patient brow reserved, as one who read
New annals in the history of man.

The blueberries which the mountain afforded, added to the milk we had brought, made our frugal supper, while for entertainment the even-song of the wood thrush rang along the ridge. Our eyes rested on no painted ceiling nor carpeted hall, but on skies of Nature's painting, and hills and forests of her embroidery. Before sunset, we rambled along the ridge to the north, while a hawk soared still above us. It was a place where gods might wander, so solemn and solitary, and removed from all contagion with the plain. As the evening came on, the haze was condensed in vapor, and the landscape became more distinctly visible, and numerous sheets of water were brought to light.

"Et jam summa procul villarum culmina fumant,
Majoresque cadunt altis de montibus umbrae."

And now the tops of the villas smoke afar off,
And the shadows fall longer from the high mountains.

As we stood on the stone tower while the sun was setting, we saw the shades of night creep gradually over the valleys of the east; and the inhabitants went into their houses, and shut their doors, while the moon silently rose up, and took possession of that part. And then the same scene was repeated on the west side, as far as the Connecticut and the Green Mountains, and the sun's rays fell on us two alone, of all New England men.

It was the night but one before the full of the moon, so bright that we could see to read distinctly by moonlight, and in the evening strolled over the summit without danger. There was, by chance, a fire blazing on Monadnock that night, which lighted up the whole western horizon, and, by making us aware of a community of mountains, made our position seem less solitary. But at length the wind drove us to the shelter of our tent, and we closed its door for the night, and fell asleep.

It was thrilling to hear the wind roar over the rocks, at intervals when we waked, for it had grown quite cold and windy. The night was, in its elements, simple even to majesty in that bleak place, — a bright moonlight and a piercing wind. It was at no time darker than twilight within the tent, and we could easily see the moon through its transparent roof as we lay; for there was the moon still above us, with Jupiter and Saturn on either hand, looking down on Wachusett, and it was a satisfaction to know that they were our fellow-travelers still, as high and out of our reach as our own destiny.

Truly the stars were given for a consolation to man. We should not know but our life were fated to be always groveling, but it is permitted to behold them, and surely they are deserving of a fair destiny. We see laws which never fail, of whose failure we never conceived; and their lamps burn all the night, too, as well as all day,— so rich and lavish is that nature which can afford this superfluity of light.

The morning twilight began as soon as the moon had set, and we arose and kindled our fire, whose blaze might have been seen for thirty miles around. As the daylight increased, it was remarkable how rapidly the wind went down. There was no dew on the summit, but coldness supplied its place. When the dawn had reached its prime, we enjoyed the view of a distinct horizon line, and could fancy ourselves at sea, and the distant hills the waves in the horizon, as seen from the deck of a vessel. The cherry-birds flitted around us, the nuthatch and flicker were heard among the bushes, the titmouse perched within a few feet, and the song of the wood thrush again rang along the ridge. At length we saw the run rise up out of the sea, and shine on Massachusetts; and from this moment the atmosphere grew more and more transparent till the time of our departure, and we began to realize the extent of the view, and how the earth, in some degree, answered to the heavens in breadth, the white villages to the constellations in the sky. There was little of the sublimity and grandeur which belong to mountain scenery, but an immense landscape to ponder on a summer's day. We could see how ample and roomy is nature. As far as the eye could reach there was little life in the landscape; the few birds that flitted past did not crowd. The travelers on the remote highways, which intersect the country on every side, had no fellow-travelers for miles, before or behind. On every side, the eye ranged over successive circles of towns, rising one above another, like the terraces of a vineyard, till they were lost in the horizon. Wachusett is, in fact, the observatory of the State. There lay Massachusetts, spread out before us in its length and breadth, like a map. There was the level horizon which told of the sea on the east and south, the well-known hills of New Hampshire on the north, and the misty summits of the Hoosac and Green Mountains, first made visible to us the evening before, blue and unsubstantial, like some bank of clouds which the morning wind would dissipate, on the northwest and west. These last distant ranges, on which the eye rests unwearied, commence with an abrupt boulder in the north, beyond the Connecticut, and travel southward, with three or four peaks dimly seen. But Monadnock, rearing its masculine front in the northwest, is the grandest feature. As we beheld it, we knew that it was the height of land between the two rivers, on this side the valley of the Merrimack, on that of the Connecticut, fluctuating with their blue seas of air, — these rival vales, already teeming with Yankee men along their respective streams, born to what destiny who shall tell? Watatic and the neighboring hills, in this State and in New Hampshire, are a continuation of the same elevated range on which we were standing. But that New Hampshire bluff, — that promontory of a State, — lowering day and night on this our State of Massachusetts, will longest haunt our dreams.

We could at length realize the place mountains occupy on the land, and how they come into the general scheme of the universe. When first we climb their summits and observe their lesser irregularities, we do not give credit to the comprehensive intelligence which shaped them; but when afterward we behold their outlines in the horizon, we confess that the hand which moulded their opposite slopes, making one to balance the other, worked round a deep centre, and was privy to the plan of the universe. So is the least part of nature in its bearings referred to all space. These lesser mountain ranges, as well as the Alleghanies, run from northeast to southwest, and parallel with these mountain streams are the more fluent rivers, answering to the general direction of the coast, the bank of the great ocean stream itself. Even the clouds, with their thin bars, fall into the same direction by preference, and such even is the course of the prevailing winds, and the migration of men and birds. A mountain chain determines many things for the statesman and philosopher. The improvements of civilization rather creep along its sides than cross its summit. How often is it a barrier to prejudice and fanaticism! In passing over these heights of land, through their thin atmosphere, the follies of the plain are refined and purified; and as many species of plants do not scale their summits, so many species of folly, no doubt, do not cross the Alleghanies; it is only the hardy mountain-plant that creeps quite over the ridge, and descends into the valley beyond.

We get a dim notion of the flight of birds, especially of such as fly high in the air, by having ascended a mountain. We can now see what landmarks mountains are to their migrations; how the Catskills and Highlands have hardly sunk to them, when Wachusett and Monadnock open a passage to the northeast; how they are guided, too, in their course by the rivers and valleys; and who knows but by the stars, as well as the mountain ranges, and not by the petty landmarks which we use. The bird whose eye takes in the Green Mountains on the one side, and the ocean on the other, need not be at a loss to find its way.

At noon we descended the mountain, and, having returned to the abodes of men, turned our faces to the east again; measuring our progress, from time to time, by the more ethereal hues which the mountain assumed. Passing swiftly through Stillwater and Sterling, as with a downward impetus, we found ourselves almost at home again in the green meadows of Lancaster, so like our own Concord, for both are watered by two streams which unite near their centres, and have many other features in common. There is an unexpected refinement about this scenery; level prairies of great extent, interspersed with elms and hop-fields and groves of trees, give it almost a classic appearance. This, it will be remembered, was the scene of Mrs. Rowlandson's capture, and of other events in the Indian wars, but from this July afternoon, and under that mild exterior, those times seemed as remote as the irruption of the Goths. They were the dark age of New England. On beholding a picture of a New England village as it then appeared, with a fair open prospect, and a light on trees and river, as if it were broad noon, we find we had not thought the sun shone in those days, or that men lived in broad daylight then. We do not imagine the sun shining on hill and valley during Philip's war, nor on the war-path of Paugus, or Standish, or Church, or Lovell, with serene summer weather, but a dim twilight or night did those events transpire in. They must have fought in the shade of their own dusky deeds.

At length, as we plodded along the dusty roads, our thoughts became as dusty as they; all thought indeed stopped, thinking broke down, or proceeded only passively in a sort of rhythmical cadence of the confused material of thought, and we found ourselves mechanically repeating some familiar measure which timed with our tread; some verse of the Robin Hood ballads, for instance, which one can recommend to travel by: —

> "Sweavens are swift, sayd lyttle John,
> As the wind blows over the hill;
> For if it be never so loud this night,
> To-morrow it may be still."

And so it went, up-hill and down, till a stone interrupted the line, when a new verse was chosen: —

> "His shoote it was but loosely shott,
> Yet flewe not the arrowe in vaine,
> For it mett one of the sheriffe's men,
> And William a Trent was slaine."

There is, however, this consolation to the most wayworn traveler, upon the dustiest road, that the path his feet describe is so perfectly symbolical of human life, — now climbing the hills, now descending into the vales.

From the summits he beholds the heavens and the horizon, from the vales he looks up to the heights again. He is treading his old lessons still, and though he may be very weary and travel-worn, it is yet sincere experience.

Leaving the Nashua, we changed our route a little, and arrived at Stillriver Village, in the western part of Harvard, just as the sun was setting. From this place, which lies to the northward, upon the western slope of the same range of hills on which we had spent the noon before, in the adjacent town, the prospect is beautiful, and the grandeur of the mountain outlines unsurpassed. There was such a repose and quiet here at this hour, as if the very hillsides were enjoying the scene; and as we passed slowly along, looking back over the country we had traversed, and listening to the evening song of the robin, we could not help contrasting the equanimity of Nature with the bustle and impatience of man. His words and actions presume always a crisis near at hand, but she is forever silent and unpretending.

And now that we have returned to the desultory life of the plain, let us endeavor to import a little of that mountain grandeur into it. We will remember within what walls we lie, and understand that this level life too has its summit, and why from the mountain-top the deepest valleys have a tinge of blue; that there is elevation in every hour, as no part of the earth is so low that the heavens may not be seen from, and we have only to stand on the summit of our hour to command an uninterrupted horizon.

We rested that night at Harvard, and the next morning, while one bent his steps to the nearer village of

Groton, the other took his separate and solitary way to the peaceful meadows of Concord; but let him not forget to record the brave hospitality of a farmer and his wife, who generously entertained him at their board, though the poor wayfarer could only congratulate the one on the continuance of hay weather, and silently accept the kindness of the other. Refreshed by this instance of generosity, no less than by the substantial viands set before him, he pushed forward with new vigor, and reached the banks of the Concord before the sun had climbed many degrees into the heavens.

A WINTER WALK

THE wind has gently murmured through the blinds, or puffed with feathery softness against the windows, and occasionally sighed like a summer zephyr lifting the leaves along, the livelong night. The meadow mouse has slept in his snug gallery in the sod, the owl has sat in a hollow tree in the depth of the swamp, the rabbit, the squirrel, and the fox have all been housed. The watch-dog has lain quiet on the hearth, and the cattle have stood silent in their stalls. The earth itself has slept, as it were its first, not its last sleep, save when some street-sign or wood-house door has faintly creaked upon its hinge, cheering forlorn nature at her midnight work, — the only sound awake 'twixt Venus and Mars, — advertising us of a remote inward warmth, a divine cheer and fellowship, where gods are met together, but where it is very bleak for men to stand. But while the earth has slumbered, all the air has been alive with feathery flakes descending, as if some northern Ceres reigned, showering her silvery grain over all the fields.

We sleep, and at length awake to the still reality of a winter morning. The snow lies warm as cotton or down upon the window-sill; the broadened sash and frosted panes admit a dim and private light, which enhances the snug cheer within. The stillness of the morning is impressive. The floor creaks under our feet as we move toward the window to look abroad through some clear space over the fields. We see the roofs stand under their snow burden. From the eaves and fences hang stalactites of snow, and in the yard stand stalagmites covering some concealed core. The trees and shrubs rear white arms to the sky on every side; and where were walls and fences, we see fantastic forms stretching in frolic gambols across the dusky landscape, as if Nature had strewn her fresh designs over the fields by night as models for man's art.

Silently we unlatch the door, letting the drift fall in, and step abroad to face the cutting air. Already the stars have lost some of their sparkle, and a dull, leaden mist skirts the horizon. A lurid brazen light in the east proclaims the approach of day, while the western landscape is dim and spectral still, and clothed in a sombre Tartarean light, like the shadowy realms. They are Infernal sounds only that you hear,—the crowing of cocks, the barking of dogs, the chopping of wood, the lowing of kine, all seem to come from Pluto's barnyard and beyond the Styx, — not for any melancholy they suggest, but their twilight bustle is too solemn and mysterious for earth. The recent tracks of the fox or otter, in the yard, remind us that each hour of the night is crowded with events, and the primeval nature is still working and making tracks in the snow. Opening the gate, we tread briskly along the lone country road, crunching the dry and crisped snow under our feet, or aroused by the sharp, clear creak of the wood-sled, just starting for the distant market, from the early farmer's door, where it has lain the summer long, dreaming amid the chips and stubble; while far through the drifts and powdered windows we see the farmer's early candle, like a paled star, emitting a lonely beam, as if some severe virtue were at its matins there. And one by one the smokes begin to ascend from the chimneys amid the trees and snows.

The sluggish smoke curls up from some deep dell,
The stiffened air exploring in the dawn,
And making slow acquaintance with the day
Delaying now upon its heavenward course,
In wreathèd loiterings dallying with itself,
With as uncertain purpose and slow deed
As its half-wakened master by the hearth,
Whose mind still slumbering and sluggish thoughts
Have not yet swept into the onward current
Of the new day; — and now it streams afar,
The while the chopper goes with step direct,
And mind intent to swing the early axe.
  First in the dusky dawn he sends abroad
His early scout, his emissary, smoke,
The earliest, latest pilgrim from the roof,
To feel the frosty air, inform the day;
And while he crouches still beside the hearth,
Nor musters courage to unbar the door,
It has gone down the glen with the light wind,
And o'er the plain unfurled its venturous wreath,
Draped the tree-tops, loitered upon the hill,
And warmed the pinions of the early bird;
And now, perchance, high in the crispy air,
Has caught sight of the day o'er the earth's edge,
And greets its master's eye at his low door,
As some refulgent cloud in the upper sky.

We hear the sound of wood-chopping at the farmers' doors, far over the frozen earth, the baying of the house-dog, and the distant clarion of the cock, — though the thin and frosty air conveys only the finer particles of sound to our ears, with short and sweet vibrations, as the waves subside soonest on the purest and lightest liquids, in which gross substances sink to the bottom. They come clear and bell-like, and from a greater distance in the horizon, as if there were fewer impediments than in summer to make them faint and ragged. The ground is sonorous, like seasoned wood, and even the ordinary rural sounds are melodious, and the jingling of the ice on the trees is sweet and liquid. There is the least possible moisture in the atmosphere, all being dried up or congealed, and it is of such extreme tenuity and elasticity that it becomes a source of delight. The withdrawn and tense sky seems groined like the aisles of a cathedral, and the polished air sparkles as if there were crystals of ice floating in it. As they who have resided in Greenland tell us that when it freezes "the sea smokes like burning turf-land, and a fog or mist arises, called frost-smoke," which "cutting smoke frequently raises blisters on the face and hands, and is very pernicious to the health." But this pure, stinging cold is an elixir to the lungs, and not so much a frozen mist as a crystallized midsummer haze, refined and purified by cold.

The sun at length rises through the distant woods, as if with the faint clashing, swinging sound of cymbals, melting the air with his beams, and with such rapid steps the morning travels, that already his rays are gilding the distant western mountains. Meanwhile we step hastily along through the powdery snow, warmed by an inward heat, enjoying an Indian summer still, in the increased glow of thought and feeling. Probably if our lives were more conformed to nature, we should not need to defend ourselves against her heats and colds, but find her our constant nurse and friend, as do plants and quadrupeds. If our bodies were fed with pure and simple elements, and not with a stimulating and heating diet, they would afford no more pasture for cold than a leafless twig, but thrive like the trees, which find even winter genial to their expansion.

The wonderful purity of nature at this season is a most pleasing fact. Every decayed stump and moss-grown stone and rail, and the dead leaves of autumn, are concealed by a clean napkin of snow. In the bare fields and tinkling woods, see what virtue survives. In the coldest and bleakest places, the warmest charities still maintain a foothold. A cold and searching wind drives away all contagion, and nothing can withstand it but what has a virtue in it, and accordingly, whatever we meet with in cold and bleak places, as the tops of mountains, we respect for a sort of sturdy innocence, a Puritan toughness. All things beside seem to be called in for shelter, and what stays out must be part of the original frame of the universe, and of such valor as God himself. It is invigorating to breathe the cleansed air. Its greater fineness and purity are visible to the eye, and we would fain stay out long and late, that the gales may sigh through us, too, as through the leafless trees, and fit us for the winter, — as if we hoped so to borrow some pure and steadfast virtue, which will stead us in all seasons.

There is a slumbering subterranean fire in nature which never goes out, and which no cold can chill. It finally melts the great snow, and in January or July is only buried under a thicker or thinner covering. In the coldest day it flows somewhere, and the snow melts around every tree. This field of winter rye, which sprouted late in the fall, and now speedily dissolves the snow, is where the fire is very thinly covered. We feel warmed by it. In the winter, warmth stands for all virtue, and we resort in thought to a trickling rill, with its bare stones shining in the sun, and to warm springs in the woods, with as much eagerness as rabbits and robins. The steam which rises from swamps and pools is as dear and domestic as that of our own kettle. What fire could ever equal the sunshine of a winter's day, when the meadow mice come out by the wall-sides, and the chickadee lisps in the defiles of the wood? The warmth comes directly from the sun, and is not radiated from the earth, as in summer; and when we feel his beams on our backs as we are treading some snowy dell, we are grateful as for a special kindness, and bless the sun which has followed us into that by-place.

This subterranean fire has its altar in each man's breast; for in the coldest day, and on the bleakest hill, the traveler cherishes a warmer fire within the folds of his cloak than is kindled on any hearth. A healthy man, indeed, is the complement of the seasons, and in winter, summer is in his heart. There is the south. Thither have all birds and insects migrated, and around the warm springs in his breast are gathered the robin and the lark.

At length, having reached the edge of the woods, and shut out the gadding town, we enter within their covert as we go under the roof of a cottage, and cross its threshold, all ceiled and banked up with snow. They are glad and warm still, and as genial and cheery in winter as in summer. As we stand in the midst of the pines in the flickering and checkered light which straggles but little way into their maze, we wonder if the towns have ever heard their simple story. It seems to us that no traveler has ever explored them, and notwithstanding the wonders which science is elsewhere revealing every day, who would not like to hear their annals? Our humble villages in the plain are their contribution. We borrow from the forest the boards which shelter and the sticks which warm us. How important is their evergreen to the winter, that portion of the summer which does not fade, the permanent year, the unwithered grass! Thus simply, and with little expense of altitude, is the surface of the earth diversified. What would human life be without forests, those natural cities? From the tops of mountains they appear like smooth-shaven lawns, yet whither shall we walk but in this taller grass?

In this glade covered with bushes of a year's growth, see how the silvery dust lies on every seared leaf and twig, deposited in such infinite and luxurious forms as by their very variety atone for the absence of color. Observe the tiny tracks of mice around every stem, and the triangular tracks of the rabbit. A pure elastic heaven hangs over all, as if the impurities of the summer sky, refined and shrunk by the chaste winter's cold, had been winnowed from the heavens upon the earth.

Nature confounds her summer distinctions at this season. The heavens seem to be nearer the earth. The elements are less reserved and distinct. Water turns to ice, rain to snow. The day is but a Scandinavian night. The winter is an arctic summer.

How much more living is the life that is in nature, the furred life which still survives the stinging nights, and, from amidst fields and woods covered with frost and snow, sees the sun rise!

> "The foodless wilds
> Pour forth their brown inhabitants."

The gray squirrel and rabbit are brisk and playful in the remote glens, even on the morning of the cold Friday. Here is our Lapland and Labrador, and for our Esquimaux and Knistenaux, Dog-ribbed Indians, Novazemblaites, and Spitzbergeners, are there not the ice-cutter and woodchopper, the fox, muskrat, and mink?

Still, in the midst of the arctic day, we may trace the summer to its retreats, and sympathize with some contemporary life. Stretched over the brooks, in the midst of the frost-bound meadows, we may observe the submarine cottages of the caddis-worms, the larvæ of the Plicipennes; their small cylindrical cases built around themselves, composed of flags, sticks, grass, and withered leaves, shells, and pebbles, in form and color like the wrecks which strew the bottom, — now drifting along over the pebbly bottom, now whirling in tiny eddies and dashing down steep falls, or sweeping rapidly along with the current, or else swaying to and fro at the end of some grass-blade or root. Anon they will leave their sunken habitations, and, crawling up the stems of plants, or to the surface, like gnats, as perfect insects henceforth, flutter over the surface of the water, or sacrifice their short lives in the flame of our candles at evening. Down yonder little glen the shrubs are drooping under their burden, and the red alderberries contrast with the white ground. Here are the marks of a myriad feet which have already been abroad. The sun rises as proudly over such a glen as over the valley of the Seine or the Tiber, and it seems the residence of a pure and self-subsistent valor, such as they never witnessed,—which never knew defeat nor fear. Here reign the simplicity and purity of a primitive age, and a health and hope far remote from towns and cities. Standing quite alone, far in the forest, while the wind is shaking down snow from the trees, and leaving the only human tracks behind us, we find our reflections of a richer variety than the life of cities. The chickadee and nuthatch are more inspiring society than statesmen and philosophers, and we shall return to these last as to more vulgar companions. In this lonely glen, with its brook draining the slopes, its creased ice and crystals of all hues, where the spruces and hemlocks stand up on either side, and the rush and sere wild oats in the rivulet itself, our lives are more serene and worthy to contemplate.

As the day advances, the heat of the sun is reflected by the hillsides, and we hear a faint but sweet music, where flows the rill released from its fetters, and the icicles are melting on the trees; and the nuthatch and partridge are heard and seen. The south wind melts the snow at noon, and the bare ground appears with its withered grass and leaves, and we are invigorated by the perfume which exhales from it, as by the scent of strong meats.

Let us go into this deserted woodman's hut, and see how he has passed the long winter nights and the short and stormy days. For here man has lived under this south hillside, and it seems a civilized and public spot. We have such associations as when the traveler stands by the ruins of Palmyra or Hecatompolis. Singing birds and flowers perchance have begun to appear here, for flowers as well as weeds follow in the footsteps of man. These hemlocks whispered over his head, these hickory logs were his fuel, and these pitch pine roots kindled his fire; yonder fuming rill in the hollow, whose thin and airy vapor still ascends as busily as ever, though he is far off now, was his well. These hemlock boughs, and the straw upon this raised platform, were his bed, and this broken dish held his drink. But he has not been here this season, for the phœbes built their nest upon this shelf last summer. I find some embers left as if he had but just gone out, where he baked his pot of beans; and while at evening he smoked his pipe, whose stemless bowl lies in the ashes, chatted with his only companion, if perchance he had any, about the depth of the snow on the morrow, already falling fast and thick without, or disputed whether the last sound was the screech of an owl, or the creak of a bough, or imagination only; and through his broad chimney-throat, in the late winter evening, ere he stretched himself upon the straw, he looked up to learn the progress of the storm, and, seeing the bright stars of Cassiopeia's Chair shining brightly down upon him, fell contentedly asleep.

See how many traces from which we may learn the chopper's history! From this stump we may guess the sharpness of his axe, and from the slope of the stroke, on which side he stood, and whether he cut down the tree without going round it or changing hands; and, from the flexure of the splinters, we may know which way it fell. This one chip contains inscribed on it the whole history of the woodchopper and of the world. On this scrap of paper, which held his sugar or salt, perchance, or was the wadding of his gun, sitting on a log in the forest, with what interest we read the tattle of cities, of those larger huts, empty and to let, like this, in High Streets and Broadways. The eaves are dripping on the south side of this simple roof, while the titmouse lisps in the pine and the genial warmth of the sun around the door is somewhat kind and human.

After two seasons, this rude dwelling does not deform the scene. Already the birds resort to it, to build their nests, and you may track to its door the feet of many quadrupeds. Thus, for a long time, nature overlooks the encroachment and profanity of man. The wood still cheerfully and unsuspiciously echoes the strokes of the axe that fells it, and while they are few and seldom, they enhance its wildness, and all the elements strive to naturalize the sound.

Now our path begins to ascend gradually to the top of this high hill, from whose precipitous south side we can look over the broad country of forest and field and river, to the distant snowy mountains. See yonder thin column of smoke curling up through the woods from some invisible farmhouse, the standard raised over some rural homestead. There must be a warmer and more genial spot there below, as where we detect the vapor from a spring forming a cloud above the trees. What fine relations are established between the traveler who discovers this airy column from some eminence in the forest and him who sits below! Up goes the smoke as silently and naturally as the vapor exhales from the leaves, and as busy disposing itself in wreaths as the housewife on the hearth below. It is a hieroglyphic of man's life, and suggests more intimate and important things than the boiling of a pot. Where its fine column rises above the forest, like an ensign, some human life has planted itself, — and such is the beginning of Rome, the establishment of the arts, and the foundation of empires, whether on the prairies of America or the steppes of Asia.

And now we descend again, to the brink of this woodland lake, which lies in a hollow of the hills, as if it were their expressed juice, and that of the leaves which are annually steeped in it. Without outlet or inlet to the eye, it has still its history, in the lapse of its waves, in the rounded pebbles on its shore, and in the pines which grow down to its brink. It has not been idle, though sedentary, but, like Abu Musa, teaches that "sitting still at home is the heavenly way; the going out is the way of the world." Yet in its evaporation it travels as far as any. In summer it is the earth's liquid eye, a mirror in the breast of nature. The sins of the wood are washed out in it. See how the woods form an amphitheatre about it, and it is an arena for all the genialness of nature. All trees direct the traveler to its brink, all paths seek it out, birds fly to it, quadrupeds flee to it, and the very ground inclines toward it. It is nature's saloon, where she has sat down to her toilet. Consider her silent economy and tidiness; how the sun comes with his evaporation to sweep the dust from its surface each morning, and a fresh surface is constantly welling up; and annually, after whatever impurities have accumulated herein, its liquid transparency appears again in the spring. In summer a hushed music seems to sweep across its surface. But now a plain sheet of snow conceals it from our eyes, except where the wind has swept the ice bare, and the sere leaves are gliding from side to side, tacking and veering on their tiny voyages. Here is one just keeled up against a pebble on shore, a dry beech leaf, rocking still, as if it would start again. A skillful engineer, methinks, might project its course since it fell from the parent stem. Here are all the elements for such a calculation. Its present position, the direction of the wind, the level of the pond, and how much more is given. In its scarred edges and veins is its log rolled up.

We fancy ourselves in the interior of a larger house. The surface of the pond is our deal table or sanded floor, and the woods rise abruptly from its edge, like the walls of a cottage. The lines set to catch pickerel through the ice look like a larger culinary preparation, and the men stand about on the white ground like pieces of forest furniture. The actions of these men, at the distance of half a mile over the ice and snow, impress us as when we read the exploits of Alexander in history. They seem not unworthy of the scenery, and as momentous as the conquest of kingdoms.

Again we have wandered through the arches of the wood, until from its skirts we hear the distant booming of ice from yonder bay of the river, as if it were moved by some other and subtler tide than oceans know. To me it has a strange sound of home, thrilling as the voice of one's distant and noble kindred. A mild summer sun shines over forest and lake, and though there is but one green leaf for many rods, yet nature enjoys a serene health. Every sound is fraught with the same mysterious assurance of health, as well now the creaking of the boughs in January, as the soft sough of the wind in July.

When Winter fringes every bough
  With his fantastic wreath,
And puts the seal of silence now
  Upon the leaves beneath;

When every stream in its penthouse
  Goes gurgling on its way,
And in his gallery the mouse
  Nibbleth the meadow hay;

Methinks the summer still is nigh,
  And lurketh underneath,
As that same meadow mouse doth lie
  Snug in that last year's heath.

And if perchance the chickadee
  Lisp a faint note anon,
The snow is summer's canopy,
  Which she herself put on.

Fair blossoms deck the cheerful trees,
  And dazzling fruits depend;
The north wind sighs a summer breeze,
  The nipping frosts to fend,

Bringing glad tidings unto me,
  The while I stand all ear,
Of a serene eternity,
  Which need not winter fear.

Out on the silent pond straightway
  The restless ice doth crack,
And pond sprites merry gambols play
  Amid the deafening rack.

Eager I hasten to the vale,
  As if I heard brave news,
How nature held high festival,
  Which it were hard to lose.

I gambol with my neighbor ice,
  And sympathizing quake,
As each new crack darts in a trice
  Across the gladsome lake.

One with the cricket in the ground,
  And fagot on the hearth,
Resounds the rare domestic sound
  Along the forest path.

Before night we will take a journey on skates along the course of this meandering river, as full of novelty to one who sits by the cottage fire all the winter's day, as if it were over the polar ice, with Captain Parry or Franklin; following the winding of the stream, now flowing amid hills, now spreading out into fair meadows, and forming a myriad coves and bays where the pine and hemlock overarch. The river flows in the rear of the towns, and we see all things from a new and wilder side. The fields and gardens come down to it with a frankness, and freedom from pretension, which they do not wear on the highway. It is the outside and edge of the earth. Our eyes are not offended by violent contrasts. The last rail of the farmer's fence is some swaying willow bough, which still preserves its freshness, and here at length all fences stop, and we no longer cross any road. We may go far up within the country now by the most retired and level road, never climbing a hill, but by broad levels ascending to the upland meadows. It is a beautiful illustration of the law of obedience, the flow of a river; the path for a sick man, a highway down which an acorn cup may float secure with its freight. Its slight occasional falls, whose precipices would not diversify the landscape, are celebrated by mist and spray, and attract the traveler from far and near. From the remote interior, its current conducts him by broad and easy steps, or by one gentler inclined plane, to the sea. Thus by an early and constant yielding to the inequalities of the ground it secures itself the easiest passage.

No domain of nature is quite closed to man at all times, and now we draw near to the empire of the fishes. Our feet glide swiftly over unfathomed depths, where in summer our line tempted the pout and perch, and where the stately pickerel lurked in the long corridors formed by the bulrushes. The deep, impenetrable marsh, where the heron waded and bittern squatted, is made pervious to our swift shoes, as if a thousand railroads had been made into it. With one impulse we are carried to the cabin of the muskrat, that earliest settler, and see him dart away under the transparent ice, like a furred fish, to his hole in the bank; and we glide rapidly over meadows where lately "the mower whet his scythe," through beds of frozen cranberries mixed with meadow-grass. We skate near to where the blackbird, the pewee, and the kingbird hung their nests over the water, and the hornets builded from the maple in the swamp. How many gay warblers, following the sun, have radiated from this nest of silver birch and thistle-down! On the swamp's outer edge was hung the supermarine village, where no foot penetrated. In this hollow tree the wood duck reared her brood, and slid away each day to forage in yonder fen.

In winter, nature is a cabinet of curiosities, full of dried specimens, in their natural order and position. The meadows and forests are a *hortus siccus.* The leaves and grasses stand perfectly pressed by the air without screw or gum, and the birds' nests are not hung on an artificial twig, but where they builded them. We go about dry-shod to inspect the summer's work in the rank swamp, and see what a growth have got the alders, the willows, and the maples; testifying to how many warm suns, and fertilizing dews and showers. See what strides their boughs took in the luxuriant summer, — and anon these dormant buds will carry them onward and upward another span into the heavens.

Occasionally we wade through fields of snow, under whose depths the river is lost for many rods, to appear again to the right or left, where we least expected; still holding on its way underneath, with a faint, stertorous, rumbling sound, as if, like the bear and marmot, it too had hibernated, and we had followed its faint summer trail to where it earthed itself in snow and ice. At first we should have thought that rivers would be empty and dry in midwinter, or else frozen solid till the spring thawed them; but their volume is not diminished even, for only a superficial cold bridges their surfaces. The thousand springs which feed the lakes and streams are flowing still. The issues of a few surface springs only are closed, and they go to swell the deep reservoirs. Nature's wells are below the frost. The summer brooks are not filled with snow-water, nor does the mower quench his thirst with that alone. The streams are swollen when the snow melts in the spring, because nature's work has been delayed, the water being turned into ice and snow, whose particles are less smooth and round, and do not find their level so soon.

Far over the ice, between the hemlock woods and snow-clad hills, stands the pickerel-fisher, his lines set in some retired cove, like a Finlander, with his arms thrust into the pouches of his dreadnaught; with dull, snowy, fishy thoughts, himself a finless fish, separated a few inches from his race; dumb, erect, and made to be enveloped in clouds and snows, like the pines on shore. In these wild scenes, men stand about in the scenery, or move deliberately and heavily, having sacrificed the sprightliness and vivacity of towns to the dumb sobriety of nature. He does not make the scenery less wild, more than the jays and muskrats, but stands there as a part of it, as the natives are represented in the voyages of early navigators, at Nootka Sound, and on the Northwest coast, with their furs about them, before they were tempted to loquacity by a scrap of iron. He belongs to the natural family of man, and is planted deeper in nature and has more root than the inhabitants of towns. Go to him, ask what luck, and you will learn that he too is a worshiper of the unseen. Hear with what sincere deference and waving gesture in his tone he speaks of the lake pickerel, which he has never seen, his primitive and ideal race of pickerel. He is connected with the shore still, as by a fish-line, and yet remembers the season when he took fish through the ice on the pond, while the peas were up in his garden at home.

But now, while we have loitered, the clouds have gathered again, and a few straggling snowflakes are beginning to descend. Faster and faster they fall, shutting out the distant objects from sight. The snow falls on every wood and field, and no crevice is forgotten; by the river and the pond, on the hill and in the valley. Quadrupeds are confined to their coverts and the birds sit upon their perches this peaceful hour. There is not so much sound as in fair weather, but silently and gradually every slope, and the gray walls and fences, and the polished ice, and the sere leaves, which were not buried before, are concealed, and the tracks of men and beasts are lost. With so little effort does nature reassert her rule and blot out the traces of men. Hear how Homer has described the same: "The snowflakes fall thick and fast on a winter's day. The winds are lulled, and the snow falls incessant, covering the tops of the mountains, and the hills, and the plains where the lotus-tree grows, and the cultivated fields, and they are falling by the inlets and shores of the foaming sea, but are silently dissolved by the waves." The snow levels all things, and infolds them deeper in the bosom of nature, as, in the slow summer, vegetation creeps up to the entablature of the temple, and the turrets of the castle, and helps her to prevail over art.

The surly night-wind rustles through the wood, and warns us to retrace our steps, while the sun goes down behind the thickening storm, and birds seek their roosts, and cattle their stalls.

> "Drooping the lab'rer ox
> Stands covered o'er with snow, and *now* demands
> The fruit of all his toil."

Though winter is represented in the almanac as an old man, facing the wind and sleet, and drawing his cloak about him, we rather think of him as a merry woodchopper, and warm-blooded youth, as blithe as summer. The unexplored grandeur of the storm keeps up the spirits of the traveler. It does not trifle with us, but has a sweet earnestness. In winter we lead a more inward life. Our hearts are warm and cheery, like cottages under drifts, whose windows and doors are half concealed, but from whose chimneys the smoke cheerfully ascends. The imprisoning drifts increase the sense of comfort which the house affords, and in the coldest days we are content to sit over the hearth and see the sky through the chimney-top, enjoying the quiet and serene life that may be had in a warm corner by the chimney-side, or feeling our pulse by listening to the low of cattle in the street, or the sound of the flail in distant barns all the long afternoon. No doubt a skillful physician could determine our health by observing how these simple and natural sounds affected us. We enjoy now, not an Oriental, but a Boreal leisure, around warm stoves and fireplaces, and watch the shadow of motes in the sunbeams.

Sometimes our fate grows too homely and familiarly serious ever to be cruel. Consider how for three months the human destiny is wrapped in furs. The good Hebrew Revelation takes no cognizance of all this cheerful snow. Is there no religion for the temperate and frigid zones? We know of no scripture which records the pure benignity of the gods on a New England winter night. Their praises have never been sung, only their wrath deprecated. The best scripture, after all, records but a meagre faith. Its saints live reserved and austere. Let a brave, devout man spend the year in the woods of Maine or Labrador, and see if the Hebrew Scriptures speak adequately to his condition and experience, from the setting in of winter to the breaking up of the ice.

Now commences the long winter evening around the farmer's hearth, when the thoughts of the indwellers travel far abroad, and men are by nature and necessity charitable and liberal to all creatures. Now is the happy resistance to cold, when the farmer reaps his reward, and thinks of his preparedness for winter, and, through the glittering panes, sees with equanimity "the mansion of the northern bear," for now the storm is over, —

"The full ethereal round,
Infinite worlds disclosing to the view,
Shines out intensely keen; and all one cope
Of starry glitter glows from pole to pole."

THE SUCCESSION OF FOREST TREES[1]

EVERY man is entitled to come to Cattle-Show, even a transcendentalist; and for my part I am more interested in the men than in the cattle. I wish to see once more those old familiar faces, whose names I do not know, which for me represent the Middlesex country, and come as near being indigenous to the soil as a white man can; the men who are not above their business, whose coats are not too black, whose shoes do not shine very much, who never wear gloves to conceal their hands. It is true, there are some queer specimens of humanity attracted to our festival, but all are welcome. I am pretty sure to meet once more that weak-minded and whimsical fellow, generally weak-bodied too, who prefers a crooked stick for a cane; perfectly useless, you would say, only *bizarre*, fit for a cabinet, like a petrified snake. A ram's horn would be as convenient, and is yet more curiously twisted. He brings that much indulged bit of the country with him, from some town's end or other, and introduces it to Concord groves, as if he had promised it so much sometime. So some, it seems to me, elect their rulers for their crookedness. But I think that a straight stick makes the best cane, and an upright man the best ruler. Or why choose a man to do plain work who is distinguished for his oddity? However, I do not know

[1] An Address read to the Middlesex Agricultural Society in Concord, September, 1860.

but you will think that they have committed this mistake who invited me to speak to you to-day.

In my capacity of surveyor, I have often talked with some of you, my employers, at your dinner-tables, after having gone round and round and behind your farming, and ascertained exactly what its limits were. Moreover, taking a surveyor's and a naturalist's liberty, I have been in the habit of going across your lots much oftener than is usual, as many of you, perhaps to your sorrow, are aware. Yet many of you, to my relief, have seemed not to be aware of it; and, when I came across you in some out-of-the-way nook of your farms, have inquired, with an air of surprise, if I were not lost, since you had never seen me in that part of the town or county before; when, if the truth were known, and it had not been for betraying my secret, I might with more propriety have inquired if *you* were not lost, since I had never seen *you* there before. I have several times shown the proprietor the shortest way out of his wood-lot.

Therefore, it would seem that I have some title to speak to you to-day; and considering what that title is, and the occasion that has called us together, I need offer no apology if I invite your attention, for the few moments that are allotted me, to a purely scientific subject.

At those dinner-tables referred to, I have often been asked, as many of you have been, if I could tell how it happened, that when a pine wood was cut down an oak one commonly sprang up, and *vice versa*. To which I have answered, and now answer, that I can tell, — that it is no mystery to me. As I am not aware that this has been clearly shown by any one, I shall lay the more stress on this point. Let me lead you back into your wood-lots again.

When, hereabouts, a single forest tree or a forest springs up naturally where none of its kind grew before, I do not hesitate to say, though in some quarters still it may sound paradoxical, that it came from a seed. Of the various ways by which trees are *known* to be propagated, — by transplanting, cuttings, and the like, — this is the only supposable one under these circumstances. No such tree has ever been known to spring from anything else. If any one asserts that it sprang from something else, or from nothing, the burden of proof lies with him.

It remains, then, only to show how the seed is transported from where it grows to where it is planted. This is done chiefly by the agency of the wind, water, and animals. The lighter seeds, as those of pines and maples, are transported chiefly by wind and water; the heavier, as acorns and nuts, by animals.

In all the pines, a very thin membrane, in appearance much like an insect's wing, grows over and around the seed, and independent of it, while the latter is being developed within its base. Indeed this is often perfectly developed, though the seed is abortive; nature being, you would say, more sure to provide the means of transporting the seed, than to provide the seed to be transported. In other words, a beautiful thin sack is woven around the seed, with a handle to it such as the wind can take hold of, and it is then committed to the wind, expressly that it may transport the seed and extend the range of the species; and this it does, as effectually as when seeds are sent by mail in a different kind of sack from the Patent Office. There is a patent office at the seat of government of the universe, whose managers are as much interested in the dispersion of seeds as anybody at Washington can be, and their operations are infinitely more extensive and regular.

There is, then, no necessity for supposing that the pines have sprung up from nothing, and I am aware that I am not at all peculiar in asserting that they come from seeds, though the mode of their propagation *by nature* has been but little attended to. They are very extensively raised from the seed in Europe, and are beginning to be here.

When you cut down an oak wood, a pine wood will not *at once* spring up there unless there are, or have been quite recently, seed-bearing pines near enough for the seeds to be blown from them. But, adjacent to a forest of pines, if you prevent other crops from growing there, you will surely have an extension of your pine forest, provided the soil is suitable.

As for the heavy seeds and nuts which are not furnished with wings, the notion is still a very common one that, when the trees which bear these spring up where none of their kind were noticed before, they have come from seeds or other principles spontaneously generated there in an unusual manner, or which have lain dormant in the soil for centuries, or perhaps been called into activity by the heat of a burning. I do not believe these assertions, and I will state some of the ways in which, according to my observation, such forests are planted and raised.

Every one of these seeds, too, will be found to be winged or legged in another fashion. Surely it is not wonderful that cherry trees of all kinds are widely dispersed, since their fruit is well known to be the favorite food of various birds. Many kinds are called bird cherries, and they appropriate many more kinds, which are not so called. Eating cherries is a bird-like employment, and unless we disperse the seeds occasionally, as they do, I shall think that the birds have the best right to them. See how artfully the seed of a cherry is placed in order that a bird may be compelled to transport it, — in the very midst of a tempting pericarp, so that the creature that would devour this must commonly take the stone also into its mouth or bill. If you ever ate a cherry, and did not make two bites of it, you must have perceived it, — right in the centre of the luscious morsel, a large earthy residuum left on the tongue. We thus take into our mouths cherry-stones as big as peas, a dozen at once, for Nature can persuade us to do almost anything when she would compass her ends. Some wild men and children instinctively swallow these, as the birds do when in a hurry, it being the shortest way to get rid of them. Thus, though these seeds are not provided with vegetable wings, Nature has impelled the thrush tribe to take them into their bills and fly away with them; and they are winged in another sense, and more effectually than the seeds of pines, for these are carried even against the wind. The consequence is, that cherry trees grow not only here but there. The same is true of a great many other seeds.

But to come to the observation which suggested these remarks. As I have said, I suspect that I can throw some light on the fact that when hereabouts a dense pine wood is cut down, oaks and other hard woods may at once take its place. I have got only to show that the acorns and nuts, provided they are grown in the neighborhood, are regularly planted in such woods; for I assert that if an oak tree has not grown within ten miles, and man has not carried acorns thither, then an oak wood will not spring up *at once*, when a pine wood is cut down.

Apparently, there were only pines there before. They are cut off, and after a year or two you see oaks and other hard woods springing up there, with scarcely a pine amid them, and the wonder commonly is, how the seed could have lain in the ground so long without decaying. But the truth is, that it has not lain in the ground so long, but is regularly planted each year by various quadrupeds and birds.

In this neighborhood, where oaks and pines are about equally dispersed, if you look through the thickest pine wood, even the seemingly unmixed pitch pine ones, you will commonly detect many little oaks, birches, and other hard woods, sprung from seeds carried into the thicket by squirrels and other animals, and also blown thither, but which are overshadowed and choked by the pines. The denser the evergreen wood, the more likely it is to be well planted with these seeds, because the planters incline to resort with their forage to the closest covert. They also carry it into birch and other woods. This planting is carried on annually, and the oldest seedlings annually die; but when the pines are cleared off, the oaks, having got just the start they want, and now secured favorable conditions, immediately spring up to trees.

The shade of a dense pine wood is more unfavorable to the springing up of pines of the same species than of oaks within it, though the former may come up abundantly when the pines are cut, if there chance to be sound seed in the ground.

But when you cut off a lot of hard wood, very often the little pines mixed with it have a similar start, for the squirrels have carried off the nuts to the pines, and not to the more open wood, and they commonly make pretty clean work of it; and moreover, if the wood was old, the sprouts will be feeble or entirely fail; to say nothing about the soil being, in a measure, exhausted for this kind of crop.

If a pine wood is surrounded by a white oak one chiefly, white oaks may be expected to succeed when the pines are cut. If it is surrounded instead by an edging of shrub oaks, then you will probably have a dense shrub oak thicket.

I have no time to go into details, but will say, in a word, that while the wind is conveying the seeds of pines into hard woods and open lands, the squirrels and other animals are conveying the seeds of oaks and walnuts into the pine woods, and thus a rotation of crops is kept up.

I affirmed this confidently many years ago, and an occasional examination of dense pine woods confirmed me in my opinion. It has long been known to observers that squirrels bury nuts in the ground, but I am not aware that any one has thus accounted for the regular succession of forests.

On the 24th of September, in 1857, as I was paddling down the Assabet, in this town, I saw a red squirrel run along the bank under some herbage, with something large in its mouth. It stopped near the foot of a hemlock, within a couple of rods of me, and, hastily pawing a hole with its fore feet, dropped its booty into it, covered it up, and retreated part way up the trunk of the tree. As I approached the shore to examine the deposit, the squirrel, descending part way, betrayed no little anxiety about its treasure, and made two or three motions to recover it before it finally retreated. Digging there, I found two green pignuts joined together, with the thick husks on, buried about an inch and a half under the reddish soil of decayed hemlock leaves, — just the right depth to plant it. In short, this squirrel was then engaged in accomplishing two objects, to wit, laying up a store of winter food for itself, and planting a hickory wood for all creation. If the squirrel was killed, or neglected its deposit, a hickory would spring up. The nearest hickory tree was twenty rods distant. These nuts were there still just fourteen days later, but were gone when I looked again, November 21st, or six weeks later still.

I have since examined more carefully several dense woods, which are said to be, and are apparently, exclusively pine, and always with the same result. For instance, I walked the same day to a small but very dense and handsome white pine grove, about fifteen rods square, in the east part of this town. The trees are large for Concord, being from ten to twenty inches in diameter, and as exclusively pine as any wood that I know. Indeed, I selected this wood because I thought it the least likely to contain anything else. It stands on an open plain or pasture, except that it adjoins another small pine wood, which has a few little oaks in it, on the southeast side. On every other side, it was at least thirty rods from the nearest woods. Standing on the edge of this grove and looking through it, for it is quite level and free from underwood, for the most part bare, red-carpeted ground, you would have said that there was not a hardwood tree in it, young or old. But on looking carefully along over its floor I discovered, though it was not till my eye had got used to the search, that, alternating with thin ferns, and small blueberry bushes, there was, not merely here and there, but as often as every five feet and with a degree of regularity, a little oak, from three to twelve inches high, and in one place I found a green acorn dropped by the base of a pine.

I confess I was surprised to find my theory so perfectly proved in this case. One of the principal agents in this planting, the red squirrels, were all the while curiously inspecting me, while I was inspecting their plantation. Some of the little oaks had been browsed by cows, which resorted to this wood for shade.

After seven or eight years, the hard woods evidently find such a locality unfavorable to their growth, the pines being allowed to stand. As an evidence of this, I observed a diseased red maple twenty-five feet long, which had been recently prostrated, though it was still covered with green leaves, the only maple in any position in the wood.

But although these oaks almost invariably die if the pines are not cut down, it is probable that they do better for a few years under their shelter than they would anywhere else.

The very extensive and thorough experiments of the English have at length led them to adopt a method of raising oaks almost precisely like this which somewhat earlier had been adopted by Nature and her squirrels here; they have simply rediscovered the value of pines as nurses for oaks. The English experimenters seem, early and generally, to have found out the importance of using trees of some kind as nurse-plants for the young oaks. I quote from Loudon what he describes as "the ultimatum on the subject of planting and sheltering oaks," — "an abstract of the practice adopted by the government officers in the national forests" of England, prepared by Alexander Milne.

At first some oaks had been planted by themselves, and others mixed with Scotch pines; "but in all cases," says Mr. Milne, "where oaks were planted actually among the pines and surrounded by them [though the soil might be inferior], the oaks were found to be much the best." "For several years past, the plan pursued has been to plant the inclosures with Scotch pines only [a tree very similar to our pitch pine], and when the pines have got to the height of five or six feet, then to put in good strong oak plants of about four or five years' growth among the pines, — not cutting away any pines at first, unless they happen to be so strong and thick as to overshadow the oaks. In about two years it becomes necessary to shred the branches of the pines, to give light and air to the oaks, and in about two or three more years to begin gradually to remove the pines altogether, taking out a certain number each year, so that, at the end of twenty or twenty-five years, not a single Scotch pine shall be left; although, for the first ten or twelve years, the plantation may have appeared to contain nothing else but pine. The advantage of this mode of planting has been found to be that the pines dry and ameliorate the soil, destroying the coarse grass and brambles which frequently choke and injure oaks; and that no mending over is necessary, as scarcely an oak so planted is found to fail."

Thus much the English planters have discovered by patient experiment, and, for aught I know, they have taken out a patent for it; but they appear not to have discovered that it was discovered before, and that they are merely adopting the method of Nature, which she long ago made patent to all. She is all the while planting the oaks amid the pines without our knowledge, and at last, instead of government officers, we send a party of woodchoppers to cut down the pines, and so rescue an oak forest, at which we wonder as if it had dropped from the skies.

As I walk amid hickories, even in August, I hear the sound of green pignuts falling from time to time, cut off by the chickaree over my head. In the fall, I notice on the ground, either within or in the neighborhood of oak woods, on all sides of the town, stout oak twigs three or four inches long, bearing half a dozen empty acorn-cups, which twigs have been gnawed off by squirrels, on both sides of the nuts, in order to make them more portable. The jays scream and the red squirrels scold while you are clubbing and shaking the chestnut trees, for they are there on the same errand, and two of a trade never agree. I frequently see a red or gray squirrel cast down a green chestnut bur, as I am going through the woods, and I used to think, sometimes, that they were cast at me. In fact, they are so busy about it, in the midst of the chestnut season, that you cannot stand long in the woods without hearing one fall. A sportsman told me that he had, the day before, —that was in the middle of October,—seen a green chestnut bur dropped on our great river meadow, fifty rods from the nearest wood, and much further from the nearest chestnut tree, and he could not tell how it came there. Occasionally, when chestnutting in midwinter, I find thirty or forty nuts in a pile, left in its gallery, just under the leaves, by the common wood mouse (*Mus leucopus*).

But especially, in the winter, the extent to which this transportation and planting of nuts is carried on is made apparent by the snow. In almost every wood, you will see where the red or gray squirrels have pawed down through the snow in a hundred places, sometimes two feet deep, and almost always directly to a nut or a pine cone, as directly as if they had started from it and bored upward, — which you and I could not have done. It would be difficult for us to find one before the snow falls. Commonly, no doubt, they had deposited them there in the fall. You wonder if they remember the localities, or discover them by the scent. The red squirrel commonly has its winter abode in the earth under a thicket of evergreens, frequently under a small clump of evergreens in the midst of a deciduous wood. If there are any nut trees which still retain their nuts standing at a distance without the wood, their paths often lead directly to and from them. We therefore need not suppose an oak standing here and there *in* the wood in order to seed it, but if a few stand within twenty or thirty rods of it, it is sufficient.

I think that I may venture to say that every white pine cone that falls to the earth naturally in this town, before opening and losing its seeds, and almost every pitch pine one that falls at all, is cut off by a squirrel, and they begin to pluck them long before they are ripe, so that when the crop of white pine cones is a small one, as it commonly is, they cut off thus almost every one of these before it fairly ripens. I think, moreover, that their design, if I may so speak, in cutting them off green, is, partly, to prevent their opening and losing their seeds, for these are the ones for which they dig through the snow, and the only white pine cones which contain anything then. I have counted in one heap, within a diameter of four feet, the cores of 239 pitch pine cones which had been cut off and stripped by the red squirrel the previous winter.

The nuts thus left on the surface, or buried just beneath it, are placed in the most favorable circumstances for germinating. I have sometimes wondered how those which merely fell on the surface of the earth got planted; but, by the end of December, I find the chestnut of the same year partially mixed with the mould, as it were, under the decaying and mouldy leaves, where there is all the moisture and manure they want, for the nuts fall fast. In a plentiful year, a large proportion of the nuts are thus covered loosely an inch deep, and are, of course, somewhat concealed from squirrels. One winter, when the crop had been abundant, I got, with the aid of a rake, many quarts of these nuts as late as the tenth of January, and though some bought at the store the same day were more than half of them mouldy, I did not find a single mouldy one among these which I picked from under the wet and mouldy leaves, where they had been snowed on once or twice. Nature knows how to pack them best. They were still plump and tender. Apparently, they do not heat there, though wet. In the spring they were all sprouting.

Loudon says that "when the nut [of the common walnut of Europe] is to be preserved through the winter for the purpose of planting in the following spring, it should be laid in a rot-heap, as soon as gathered, with the husk on, and the heap should be turned over frequently in the course of the winter."

Here, again, he is stealing Nature's "thunder." How can a poor mortal do otherwise? for it is she that finds fingers to steal with, and the treasure to be stolen. In the planting of the seeds of most trees, the best gardeners do no more than follow Nature, though they may not know it. Generally, both large and small ones are most sure to germinate, and succeed best, when only beaten into the earth with the back of a spade, and then covered with leaves or straw. These results to which planters have arrived remind us of the experience of Kane and his companions at the north, who, when learning to live in that climate, were surprised to find themselves steadily adopting the customs of the natives, simply becoming Esquimaux. So, when we experiment in planting forests, we find ourselves at last doing as Nature does.

Would it not be well to consult with Nature in the outset? for she is the most extensive and experienced planter of us all, not excepting the Dukes of Athol.

In short, they who have not attended particularly to this subject are but little aware to what an extent quadrupeds and birds are employed, especially in the fall, in collecting, and so disseminating and planting, the seeds of trees. It is the almost constant employment of the squirrels at that season, and you rarely meet with one that has not a nut in its mouth, or is not just going to get one. One squirrel-hunter of this town told me that he knew of a walnut tree which bore particularly good nuts, but that on going to gather them one fall, he found that he had been anticipated by a family of a dozen red squirrels. He took out of the tree, which was hollow, one bushel and three pecks by measurement, without the husks, and they supplied him and his family for the winter. It would be easy to multiply instances of this kind. How commonly in the fall you see the cheek-pouches of the striped squirrel distended by a quantity of nuts! This species gets its scientific name, *Tamias*, or the steward, from its habit of storing up nuts and other seeds. Look under a nut tree a month after the nuts have fallen, and see what proportion of sound nuts to the abortive ones and shells you will find ordinarily. They have been already eaten, or dispersed far and wide. The ground looks like a platform before a grocery, where the gossips of the village sit to crack nuts and less savory jokes. You have come, you would say, after the feast was over, and are presented with the shells only.

Occasionally, when threading the woods in the fall, you will hear a sound as if some one had broken a twig, and, looking up, see a jay pecking at an acorn, or you will see a flock of them at once about it, in the top of an oak, and hear them break them off. They then fly to a suitable limb, and placing the acorn under one foot, hammer away at it busily, making a sound like a woodpecker's tapping, looking round from time to time to see if any foe is approaching, and soon reach the meat, and nibble at it, holding up their heads to swallow, while they hold the remainder very firmly with their claws. Nevertheless it often drops to the ground before the bird has done with it. I can confirm what William Bartram wrote to Wilson, the ornithologist, that "the jay is one of the most useful agents in the economy of nature, for disseminating forest trees and other nuciferous and hard-seeded vegetables on which they feed. Their chief employment during the autumnal season is foraging to supply their winter stores. In performing this necessary duty they drop abundance of seed in their flight over fields, hedges, and by fences, where they alight to deposit them in the post-holes, etc. It is remarkable what numbers of young trees rise up in fields and pastures after a wet winter and spring. These birds alone are capable, in a few years' time, to replant all the cleared lands."

I have noticed that squirrels also frequently drop their nuts in open land, which will still further account for the oaks and walnuts which spring up in pastures, for, depend on it, every new tree comes from a seed. When I examine the little oaks, one or two years old, in such places, I invariably find the empty acorn from which they sprung.

So far from the seed having lain dormant in the soil since oaks grew there before, as many believe, it is well known that it is difficult to preserve the vitality of acorns long enough to transport them to Europe; and it is recommended in Loudon's "Arboretum," as the safest course, to sprout them in pots on the voyage. The same authority states that "very few acorns of any species will germinate after having been kept a year," that beech mast "only retains its vital properties one year," and the black walnut "seldom more than six months after it has ripened." I have frequently found that in November almost every acorn left on the ground had sprouted or decayed. What with frost, drouth, moisture, and worms, the greater part are soon destroyed. Yet it is stated by one botanical writer that "acorns that have lain for centuries, on being ploughed up, have soon vegetated."

Mr. George B. Emerson, in his valuable Report on the Trees and Shrubs of this State, says of the pines: "The tenacity of life of the seeds is remarkable. They will remain for many years unchanged in the ground, protected by the coolness and deep shade of the forest above them. But when the forest is removed, and the warmth of the sun admitted, they immediately vegetate." Since he does not tell us on what observation his remark is founded, I must doubt its truth. Besides, the experience of nursery-men makes it the more questionable.

The stories of wheat raised from seed buried with an ancient Egyptian, and of raspberries raised from seed found in the stomach of a man in England, who is supposed to have died sixteen or seventeen hundred years ago, are generally discredited, simply because the evidence is not conclusive.

Several men of science, Dr. Carpenter among them, have used the statement that beach plums sprang up in sand which was dug up forty miles inland in Maine, to prove that the seed had lain there a very long time, and some have inferred that the coast has receded so far. But it seems to me necessary to their argument to show, first, that beach plums grow only on a beach. They are not uncommon here, which is about half that distance from the shore; and I remember a dense patch a few miles north of us, twenty-five miles inland, from which the fruit was annually carried to market. How much further inland they grow, I know not. Dr. Charles T. Jackson speaks of finding "beach plums" (perhaps they were this kind) more than one hundred miles inland in Maine.

It chances that similar objections lie against all the more notorious instances of the kind on record.

Yet I am prepared to believe that some seeds, especially small ones, may retain their vitality for centuries under favorable circumstances. In the spring of 1859, the old Hunt house, so called, in this town, whose chimney bore the date 1703, was taken down. This stood on land which belonged to John Winthrop, the first governor of Massachusetts, and a part of the house was evidently much older than the above date, and belonged to the Winthrop family. For many years I have ransacked this neighborhood for plants, and I consider myself familiar with its productions. Thinking of the seeds which are said to be sometimes dug up at an unusual depth in the earth, and thus to reproduce long extinct plants, it occurred to me last fall that some new or rare plants might have sprung up in the cellar of this house, which had been covered from the light so long. Searching there on the 22d of September, I found, among other rank weeds, a species of nettle (*Urtica urens*) which I had not found before; dill, which I had not seen growing spontaneously; the Jerusalem oak (*Chenopodium Botrys*), which I had seen wild in but one place; black nightshade (*Solanum nigrum*), which is quite rare hereabouts, and common tobacco, which, though it was often cultivated here in the last century, has for fifty years been an unknown plant in this town, and a few months before this not even I had heard that one man, in the north part of the town, was cultivating a few plants for his own use. I have no doubt that some or all of these plants sprang from seeds which had long been buried under or about that house, and that that tobacco is an additional evidence that the plant was formerly cultivated here. The cellar has been filled up this year, and four of those plants, including the tobacco, are now again extinct in that locality.

It is true, I have shown that the animals consume a great part of the seeds of trees, and so, at least, effectually prevent their becoming trees; but in all these cases, as I have said, the consumer is compelled to be at the same time the disperser and planter, and this is the tax which he pays to Nature. I think it is Linnæus who says that while the swine is rooting for acorns he is planting acorns.

Though I do not believe that a plant will spring up where no seed has been, I have great faith in a seed,—a, to me, equally mysterious origin for it. Convince me that you have a seed there, and I am prepared to expect wonders. I shall even believe that the millennium is at hand, and that the reign of justice is about to commence, when the Patent Office, or Government, begins to distribute, and the people to plant, the seeds of these things.

In the spring of 1857 I planted six seeds sent to me from the Patent Office, and labeled, I think, *Poitrine jaune grosse*, large yellow squash. Two came up, and one bore a squash which weighed 123½ pounds, the other bore four, weighing together 186¼ pounds. Who would have believed that there was 310 pounds of *poitrine jaune grosse* in that corner of my garden? These seeds were the bait I used to catch it, my ferrets which I sent into its burrow, my brace of terriers which unearthed it. A little mysterious hoeing and manuring was all the *abracadabra presto-change* that I used, and lo! true to the label, they found for me 310 pounds of *poitrine jaune grosse* there, where it never was known to be, nor was before. These talismans had perchance sprung from America at first, and returned to it with unabated force. The big squash took a premium at your fair that fall, and I understood that the man who bought it, intended to sell the seeds for ten cents apiece. (Were they not cheap at that?) But I have more hounds of the same breed. I learn that one which I despatched to a distant town, true to its instincts, points to the large yellow squash there, too, where no hound ever found it before, as its ancestors did here and in France.

Other seeds I have which will find other things in that corner of my garden, in like fashion, almost any fruit you wish, every year for ages, until the crop more than fills the whole garden. You have but little more to do than throw up your cap for entertainment these American days. Perfect alchemists I keep who can transmute substances without end, and thus the corner of my garden is an inexhaustible treasure-chest. Here you can dig, not gold, but the value which gold merely represents; and there is no Signor Blitz about it. Yet farmers' sons will stare by the hour to see a juggler draw ribbons from his throat, though he tells them it is all deception. Surely, men love darkness rather than light.

WALKING

I WISH to speak a word for Nature, for absolute freedom and wildness, as contrasted with a freedom and culture merely civil, — to regard man as an inhabitant, or a part and parcel of Nature, rather than a member of society. I wish to make an extreme statement, if so I may make an emphatic one, for there are enough champions of civilization: the minister and the school committee and every one of you will take care of that.

I have met with but one or two persons in the course of my life who understood the art of Walking, that is, of taking walks, — who had a genius, so to speak, for *sauntering*, which word is beautifully derived "from idle people who roved about the country, in the Middle Ages, and asked charity, under pretense of going *à la Sainte Terre*," to the Holy Land, till the children exclaimed, "There goes a *Sainte-Terrer*," a Saunterer, a Holy-Lander. They who never go to the Holy Land in their walks, as they pretend, are indeed mere idlers and vagabonds; but they who do go there are saunterers in the good sense, such as I mean. Some, however, would derive the word from *sans terre*, without land or a home, which, therefore, in the good sense, will mean, having no particular home, but equally at home everywhere. For this is the secret of successful sauntering. He who sits still in a house all the time may be the greatest vagrant of all; but the saunterer, in the good sense, is no more vagrant than the meandering river, which is all the while sedulously seeking the shortest course to the sea. But I prefer the first, which, indeed, is the most probable derivation. For every walk is a sort of crusade, preached by some Peter the Hermit in us, to go forth and reconquer this Holy Land from the hands of the Infidels.

It is true, we are but faint-hearted crusaders, even the walkers, nowadays, who undertake no persevering, never-ending enterprises. Our expeditions are but tours, and come round again at evening to the old hearth-side from which we set out. Half the walk is but retracing our steps. We should go forth on the shortest walk, perchance, in the spirit of undying adventure, never to return, — prepared to send back our embalmed hearts only as relics to our desolate kingdoms. If you are ready to leave father and mother, and brother and sister, and wife and child and friends, and never see them again, — if you have paid your debts, and made your will, and settled all your affairs, and are a free man, then you are ready for a walk.

To come down to my own experience, my companion and I, for I sometimes have a companion, take pleasure in fancying ourselves knights of a new, or rather an old, order, — not Equestrians or Chevaliers, not Ritters or Riders, but Walkers, a still more ancient and honorable class, I trust. The chivalric and heroic spirit which once belonged to the Rider seems now to reside in, or perchance to have subsided into, the Walker, — not the Knight, but Walker, Errant. He is a sort of fourth estate, outside of Church and State and People.

We have felt that we almost alone hereabouts practiced this noble art; though, to tell the truth, at least if their own assertions are to be received, most of my townsmen would fain walk sometimes, as I do, but they cannot. No wealth can buy the requisite leisure, freedom, and independence which are the capital in this profession. It comes only by the grace of God. It requires a direct dispensation from Heaven to become a walker. You must be born into the family of the Walkers. *Ambulator nascitur, non fit.* Some of my townsmen, it is true, can remember and have described to me some walks which they took ten years ago, in which they were so blessed as to lose themselves for half an hour in the woods; but I know very well that they have confined themselves to the highway ever since, whatever pretensions they may make to belong to this select class. No doubt they were elevated for a moment as by the reminiscence of a previous state of existence, when even they were foresters and outlaws.

> "When he came to grene wode,
> In a mery mornynge,
> There he herde the notes small
> Of byrdes mery syngynge.
>
> "It is ferre gone, sayd Robyn,
> That I was last here;
> Me lyste a lytell for to shote
> At the donne dere."

I think that I cannot preserve my health and spirits, unless I spend four hours a day at least — and it is commonly more than that — sauntering through the woods and over the hills and fields, absolutely free from all worldly engagements. You may safely say, A penny for your thoughts, or a thousand pounds. When sometimes I am reminded that the mechanics and shopkeepers stay in their shops not only all the forenoon, but all the afternoon too, sitting with crossed legs, so many of them, — as if the legs were made to sit upon, and not to stand or walk upon, — I think that they deserve some credit for not having all committed suicide long ago.

I, who cannot stay in my chamber for a single day without acquiring some rust, and when sometimes I have stolen forth for a walk at the eleventh hour, or four o'clock in the afternoon, too late to redeem the day, when the shades of night were already beginning to be mingled with the daylight, have felt as if I had committed some sin to be atoned for, — I confess that I am astonished at the power of endurance, to say nothing of the moral insensibility, of my neighbors who confine themselves to shops and offices the whole day for weeks and months, aye, and years almost together. I know not what manner of stuff they are of, — sitting there now at three o'clock in the afternoon, as if it were three o'clock in the morning. Bonaparte may talk of the three-o'clock-in-the-morning courage, but it is nothing to the courage which can sit down cheerfully at this hour in the afternoon over against one's self whom you have known all the morning, to starve out a garrison to whom you are bound by such strong ties of sympathy. I wonder that about this time, or say between four and five o'clock in the afternoon, too late for the morning papers and too early for the evening ones, there is not a general explosion heard up and down the street, scattering a legion of antiquated and house-bred notions and whims to the four winds for an airing, — and so the evil cure itself.

How womankind, who are confined to the house still more than men, stand it I do not know; but I have ground to suspect that most of them do not *stand* it at all. When, early in a summer afternoon, we have been shaking the dust of the village from the skirts of our garments, making haste past those houses with purely Doric or Gothic fronts, which have such an air of repose about them, my companion whispers that probably about these times their occupants are all gone to bed. Then it is that I appreciate the beauty and the glory of architecture, which itself never turns in, but forever stands out and erect, keeping watch over the slumberers.

No doubt temperament, and, above all, age, have a good deal to do with it. As a man grows older, his ability to sit still and follow indoor occupations increases. He grows vespertinal in his habits as the evening of life approaches, till at last he comes forth only just before sundown, and gets all the walk that he requires in half an hour.

But the walking of which I speak has nothing in it akin to taking exercise, as it is called, as the sick take medicine at stated hours, — as the swinging of dumb-bells or chairs; but is itself the enterprise and adventure of the day. If you would get exercise, go in search of the springs of life. Think of a man's swinging dumb-bells for his health, when those springs are bubbling up in far-off pastures unsought by him!

Moreover, you must walk like a camel, which is said to be the only beast which ruminates when walking. When a traveler asked Wordsworth's servant to show him her master's study, she answered, "Here is his library, but his study is out of doors."

Living much out of doors, in the sun and wind, will no doubt produce a certain roughness of character, — will cause a thicker cuticle to grow over some of the finer qualities of our nature, as on the face and hands, or as severe manual labor robs the hands of some of their delicacy of touch. So staying in the house, on the other hand, may produce a softness and smoothness, not to say thinness of skin, accompanied by an increased sensibility to certain impressions. Perhaps we should be more susceptible to some influences important to our intellectual and moral growth, if the sun had shone and the wind blown on us a little less; and no doubt it is a nice matter to proportion rightly the thick and thin skin. But methinks that is a scurf that will fall off fast enough, — that the natural remedy is to be found in the proportion which the night bears to the day, the winter to the summer, thought to experience. There will be so much the more air and sunshine in our thoughts. The callous palms of the laborer are conversant with finer tissues of self-respect and heroism, whose touch thrills the heart, than the languid fingers of idleness. That is mere sentimentality that lies abed by day and thinks itself white, far from the tan and callus of experience.

When we walk, we naturally go to the fields and woods: what would become of us, if we walked only in a garden or a mall? Even some sects of philosophers have felt the necessity of importing the woods to themselves, since they did not go to the woods. "They planted groves and walks of Platanes," where they took *subdiales ambulationes* in porticos open to the air. Of course it is of no use to direct our steps to the woods, if they do not carry us thither. I am alarmed when it happens that I have walked a mile into the woods bodily, without getting there in spirit. In my afternoon walk I would fain forget all my morning occupations and my obligations to society. But it sometimes happens that I cannot easily shake off the village. The thought of some work will run in my head and I am not where my body is, — I am out of my senses. In my walks I would fain return to my senses. What business have I in the woods, if I am thinking of something out of the woods? I suspect myself, and cannot help a shudder, when I find myself so implicated even in what are called good works, — for this may sometimes happen.

My vicinity affords many good walks; and though for so many years I have walked almost every day, and sometimes for several days together, I have not yet exhausted them. An absolutely new prospect is a great happiness, and I can still get this any afternoon. Two or three hours' walking will carry me to as strange a country as I expect ever to see. A single farmhouse which I had not seen before is sometimes as good as the dominions of the King of Dahomey. There is in fact a sort of harmony discoverable between the capabilities of the landscape within a circle of ten miles' radius, or the limits of an afternoon walk, and the threescore years and ten of human life. It will never become quite familiar to you.

Nowadays almost all man's improvements, so called, as the building of houses and the cutting down of the forest and of all large trees, simply deform the landscape, and make it more and more tame and cheap. A people who would begin by burning the fences and let the forest stand! I saw the fences half consumed, their ends lost in the middle of the prairie, and some worldly miser with a surveyor looking after his bounds, while heaven had taken place around him, and he did not see the angels going to and fro, but was looking for an old post-hole in the midst of paradise. I looked again, and saw him standing in the middle of a boggy Stygian fen, surrounded by devils, and he had found his bounds without a doubt, three little stones, where a stake had been driven, and looking nearer, I saw that the Prince of Darkness was his surveyor.

I can easily walk ten, fifteen, twenty, any number of miles, commencing at my own door, without going by any house, without crossing a road except where the fox and the mink do: first along by the river, and then the brook, and then the meadow and the woodside. There are square miles in my vicinity which have no inhabitant. From many a hill I can see civilization and the abodes of man afar. The farmers and their works are scarcely more obvious than woodchucks and their burrows. Man and his affairs, church and state and school, trade and commerce, and manufactures and agriculture, even politics, the most alarming of them all, — I am pleased to see how little space they occupy in the landscape. Politics is but a narrow field, and that still narrower highway yonder leads to it. I sometimes direct the traveler thither. If you would go to the political world, follow the great road, — follow that market-man, keep his dust in your eyes, and it will lead you straight to it; for it, too, has its place merely, and does not occupy all space. I pass from it as from a bean-field into the forest, and it is forgotten. In one half-hour I can walk off to some portion of the earth's surface where a man does not stand from one year's end to another, and there, consequently, politics are not, for they are but as the cigar-smoke of a man.

The village is the place to which the roads tend, a sort of expansion of the highway, as a lake of a river. It is the body of which roads are the arms and legs, — a trivial or quadrivial place, the thoroughfare and ordinary of travelers. The word is from the Latin *villa*, which together with *via*, a way, or more anciently *ved* and *vella*, Varro derives from *veho*, to carry, because the villa is the place to and from which things are carried. They who got their living by teaming were said *vellaturam facere*. Hence, too, the Latin word *vilis* and our vile, also *villain*. This suggests what kind of degeneracy villagers are liable to. They are wayworn by the travel that goes by and over them, without traveling themselves.

Some do not walk at all; others walk in the highways; a few walk across lots. Roads are made for horses and men of business. I do not travel in them much, comparatively, because I am not in a hurry to get to any tavern or grocery or livery-stable or depot to which they lead. I am a good horse to travel, but not from choice a roadster. The landscape-painter uses the figures of men to mark a road. He would not make that use of my figure. I walk out into a nature such as the old prophets and poets, Menu, Moses, Homer, Chaucer, walked in. You may name it America, but it is not America; neither Americus Vespucius, nor Columbus, nor the rest were the discoverers of it. There is a truer account of it in mythology than in any history of America, so called, that I have seen.

However, there are a few old roads that may be trodden with profit, as if they led somewhere now that they are nearly discontinued. There is the Old Marlborough Road, which does not go to Marlborough now, methinks, unless that is Marlborough where it carries me. I am the bolder to speak of it here, because I presume that there are one or two such roads in every town.

THE OLD MARLBOROUGH ROAD

Where they once dug for money,
But never found any;
Where sometimes Martial Miles
Singly files,
And Elijah Wood,
I fear for no good:
No other man,
Save Elisha Dugan, —
O man of wild habits,
Partridges and rabbits,
Who hast no cares
Only to set snares,
Who liv'st all alone,
Close to the bone,

And where life is sweetest
Constantly eatest.
When the spring stirs my blood
With the instinct to travel,
I can get enough gravel
On the Old Marlborough Road.
Nobody repairs it,
For nobody wears it;
It is a living way,
As the Christians say.
Not many there be
Who enter therein,
Only the guests of the
Irishman Quin.
What is it, what is it,
But a direction out there,
And the bare possibility
Of going somewhere?
Great guide-boards of stone,
But travelers none;
Cenotaphs of the towns
Named on their crowns.
It is worth going to see
Where you *might* be.
What king
Did the thing,
I am still wondering;
Set up how or when,
By what selectmen,
Gourgas or Lee,
Clark or Darby?
They 're a great endeavor
To be something forever;
Blank tablets of stone,
Where a traveler might groan,
And in one sentence
Grave all that is known;
Which another might read,
In his extreme need.

I know one or two
Lines that would do,
Literature that might stand
All over the land,
Which a man could remember
Till next December,
And read again in the spring,
After the thawing.
If with fancy unfurled
You leave your abode,
You may go round the world
By the Old Marlborough Road.

At present, in this vicinity, the best part of the land is not private property; the landscape is not owned, and the walker enjoys comparative freedom. But possibly the day will come when it will be partitioned off into so-called pleasure-grounds, in which a few will take a narrow and exclusive pleasure only,—when fences shall be multiplied, and man-traps and other engines invented to confine men to the *public* road, and walking over the surface of God's earth shall be construed to mean trespassing on some gentleman's grounds. To enjoy a thing exclusively is commonly to exclude yourself from the true enjoyment of it. Let us improve our opportunities, then, before the evil days come.

What is it that makes it so hard sometimes to determine whither we will walk? I believe that there is a subtle magnetism in Nature, which, if we unconsciously yield to it, will direct us aright. It is not indifferent to us which way we walk. There is a right way; but we are very liable from heedlessness and stupidity to take the wrong one. We would fain take that walk, never yet taken by us through this actual world, which is perfectly symbolical of the path which we love to travel in the interior and ideal world; and sometimes, no doubt, we find it difficult to choose our direction, because it does not yet exist distinctly in our idea.

When I go out of the house for a walk, uncertain as yet whither I will bend my steps, and submit myself to my instinct to decide for me, I find, strange and whimsical as it may seem, that I finally and inevitably settle southwest, toward some particular wood or meadow or deserted pasture or hill in that direction. My needle is slow to settle,—varies a few degrees, and does not always point due southwest, it is true, and it has good authority for this variation, but it always settles between west and south-southwest. The future lies that way to me, and the earth seems more unexhausted and richer on that side. The outline which would bound my walks would be, not a circle, but a parabola, or rather like one of those cometary orbits which have been thought to be non-returning curves, in this case opening westward, in which my house occupies the place of the sun. I turn round and round irresolute sometimes for a quarter of an hour, until I decide, for a thousandth time, that I will walk into the southwest or west. Eastward I go only by force; but westward I go free. Thither no business leads me. It is hard for me to believe that I shall find fair landscapes or sufficient wildness and freedom behind the eastern horizon. I am not excited by the prospect of a walk thither; but I believe that the forest which I see in the western horizon stretches uninterruptedly toward the setting sun, and there are no towns nor cities in it of enough consequence to disturb me. Let me live where I will, on this side is the city, on that the wilderness, and ever I am leaving the city more and more, and withdrawing into the wilderness. I should not lay so much stress on this fact, if I did not believe that something like this is the prevailing tendency of my countrymen. I must walk toward Oregon, and not toward Europe. And that way the nation is moving, and I may say that mankind progress from east to west. Within a few years we have witnessed the phenomenon of a southeastward migration, in the settlement of Australia; but this affects us as a retrograde movement, and, judging from the moral and physical character of the first generation of Australians, has not yet proved a successful experiment. The eastern Tartars think that there is nothing west beyond Thibet. "The world ends there," say they; "beyond there is nothing but a shoreless sea." It is unmitigated East where they live.

We go eastward to realize history and study the works of art and literature, retracing the steps of the race; we go westward as into the future, with a spirit of enterprise and adventure. The Atlantic is a Lethean stream, in our passage over which we have had an opportunity to forget the Old World and its institutions. If we do not succeed this time, there is perhaps one more chance for the race left before it arrives on the banks of the Styx; and that is in the Lethe of the Pacific, which is three times as wide.

I know not how significant it is, or how far it is an evidence of singularity, that an individual should thus consent in his pettiest walk with the general movement of the race; but I know that something akin to the migratory instinct in birds and quadrupeds, — which, in some instances, is known to have affected the squirrel tribe, impelling them to a general and mysterious movement, in which they were seen, say some, crossing the broadest rivers, each on its particular chip, with its tail raised for a sail, and bridging narrower streams with their dead, — that something like the *furor* which affects the domestic cattle in the spring, and which is referred to a worm in their tails, affects both nations and individuals, either perennially or from time to time. Not a flock of wild geese cackles over our town, but it to some extent unsettles the value of real estate here, and, if I were a broker, I should probably take that disturbance into account.

> "Than longen folk to gon on pilgrimages,
> And palmeres for to seken strange strondes."

Every sunset which I witness inspires me with the desire to go to a West as distant and as fair as that into which the sun goes down. He appears to migrate westward daily, and tempt us to follow him. He is the Great Western Pioneer whom the nations follow. We dream all night of those mountain-ridges in the horizon, though they may be of vapor only, which were last gilded by his rays. The island of Atlantis, and the islands and gardens of the Hesperides, a sort of terrestrial paradise, appear to have been the Great West of the ancients, enveloped in mystery and poetry. Who has not seen in imagination, when looking into the sunset sky, the gardens of the Hesperides, and the foundation of all those fables?

Columbus felt the westward tendency more strongly

than any before. He obeyed it, and found a New World for Castile and Leon. The herd of men in those days scented fresh pastures from afar.

> "And now the sun had stretched out all the hills,
> And now was dropped into the western bay;
> At last *he* rose, and twitched his mantle blue;
> To-morrow to fresh woods and pastures new."

Where on the globe can there be found an area of equal extent with that occupied by the bulk of our States, so fertile and so rich and varied in its productions, and at the same time so habitable by the European, as this is? Michaux, who knew but part of them, says that "the species of large trees are much more numerous in North America than in Europe; in the United States there are more than one hundred and forty species that exceed thirty feet in height; in France there are but thirty that attain this size." Later botanists more than confirm his observations. Humboldt came to America to realize his youthful dreams of a tropical vegetation, and he beheld it in its greatest perfection in the primitive forests of the Amazon, the most gigantic wilderness on the earth, which he has so eloquently described. The geographer Guyot, himself a European, goes farther, — farther than I am ready to follow him; yet not when he says: "As the plant is made for the animal, as the vegetable world is made for the animal world, America is made for the man of the Old World. . . . The man of the Old World sets out upon his way. Leaving the highlands of Asia, he descends from station to station towards Europe. Each of his steps is marked by a new civilization superior to the preceding, by a greater power of development. Arrived at the Atlantic, he pauses on the shore of this unknown ocean, the bounds of which he knows not, and turns upon his footprints for an instant." When he has exhausted the rich soil of Europe, and reinvigorated himself, "then recommences his adventurous career westward as in the earliest ages." So far Guyot.

From this western impulse coming in contact with the barrier of the Atlantic sprang the commerce and enterprise of modern times. The younger Michaux, in his "Travels West of the Alleghanies in 1802," says that the common inquiry in the newly settled West was, "'From what part of the world have you come?' As if these vast and fertile regions would naturally be the place of meeting and common country of all the inhabitants of the globe."

To use an obsolete Latin word, I might say, *Ex Oriente lux; ex Occidente* FRUX. From the East light; from the West fruit.

Sir Francis Head, an English traveler and a Governor-General of Canada, tells us that "in both the northern and southern hemispheres of the New World, Nature has not only outlined her works on a larger scale, but has painted the whole picture with brighter and more costly colors than she used in delineating and in beautifying the Old World. . . . The heavens of America appear infinitely higher, the sky is bluer, the air is fresher, the cold is intenser, the moon looks larger, the stars are brighter, the thunder is louder, the lightning is vivider, the wind is stronger, the rain is heavier, the mountains are higher, the rivers longer, the forests bigger, the plains broader." This statement will do at least to set against Buffon's account of this part of the world and its productions.

Linnæus said long ago, "Nescio quae facies *laeta, glabra* plantis Americanis" (I know not what there is of joyous and smooth in the aspect of American plants); and I think that in this country there are no, or at most very few, *Africanae bestiae*, African beasts, as the Romans called them, and that in this respect also it is peculiarly fitted for the habitation of man. We are told that within three miles of the centre of the East-Indian city of Singapore, some of the inhabitants are annually carried off by tigers; but the traveler can lie down in the woods at night almost anywhere in North America without fear of wild beasts.

These are encouraging testimonies. If the moon looks larger here than in Europe, probably the sun looks larger also. If the heavens of America appear infinitely higher, and the stars brighter, I trust that these facts are symbolical of the height to which the philosophy and poetry and religion of her inhabitants may one day soar. At length, perchance, the immaterial heaven will appear as much higher to the American mind, and the intimations that star it as much brighter. For I believe that climate does thus react on man, — as there is something in the mountain air that feeds the spirit and inspires. Will not man grow to greater perfection intellectually as well as physically under these influences? Or is it unimportant how many foggy days there are in his life? I trust that we shall be more imaginative, that our thoughts will be clearer, fresher, and more ethereal, as our sky, — our understanding more comprehensive and broader, like our plains, — our intellect generally on a grander scale, like our thunder and lightning, our rivers and mountains and forests, — and our hearts shall even correspond in breadth and depth and grandeur to our inland seas. Perchance there will appear to the traveler something, he knows not what, of *laeta* and *glabra*, of joyous and serene, in our very faces. Else to what end does the world go on, and why was America discovered?

To Americans I hardly need to say, —

> "Westward the star of empire takes its way."

As a true patriot, I should be ashamed to think that Adam in paradise was more favorably situated on the whole than the backwoodsman in this country.

Our sympathies in Massachusetts are not confined to New England; though we may be estranged from the South, we sympathize with the West. There is the home of the younger sons, as among the Scandinavians they took to the sea for their inheritance. It is too late to be studying Hebrew; it is more important to understand even the slang of to-day.

Some months ago I went to see a panorama of the Rhine. It was like a dream of the Middle Ages. I floated down its historic stream in something more than imagination, under bridges built by the Romans, and repaired by later heroes, past cities and castles whose very names were music to my ears, and each of which was the subject of a legend. There were Ehrenbreitstein and Rolandseck and Coblentz, which I knew only in history. They were ruins that interested me chiefly. There seemed to come up from its waters and its vine-clad hills and valleys a hushed music as of Crusaders departing for the Holy Land. I floated along under the spell of enchantment, as if I had been transported to an heroic age, and breathed an atmosphere of chivalry.

Soon after, I went to see a panorama of the Mississippi, and as I worked my way up the river in the light of to-day, and saw the steamboats wooding up, counted the rising cities, gazed on the fresh ruins of Nauvoo, beheld the Indians moving west across the stream, and, as before I had looked up the Moselle, now looked up the Ohio and the Missouri and heard the legends of Dubuque and of Wenona's Cliff, — still thinking more of the future than of the past or present, — I saw that this was a Rhine stream of a different kind; that the foundations of castles were yet to be laid, and the famous bridges were yet to be thrown over the river; and I felt that *this was the heroic age itself*, though we know it not, for the hero is commonly the simplest and obscurest of men.

The West of which I speak is but another name for the Wild; and what I have been preparing to say is, that in Wildness is the preservation of the World. Every tree sends its fibres forth in search of the Wild. The cities import it at any price. Men plow and sail for it. From the forest and wilderness come the tonics and barks which brace mankind. Our ancestors were savages. The story of Romulus and Remus being suckled by a wolf is not a meaningless fable. The founders of every state which has risen to eminence have drawn their nourishment and vigor from a similar wild source. It was because the children of the Empire were not suckled by the wolf that they were conquered and displaced by the children of the northern forests who were.

I believe in the forest, and in the meadow, and in the night in which the corn grows. We require an infusion of hemlock spruce or arbor-vitæ in our tea. There is a difference between eating and drinking for strength and from mere gluttony. The Hottentots eagerly devour the marrow of the koodoo and other antelopes raw, as a matter of course. Some of our northern Indians eat raw the marrow of the Arctic reindeer, as well as various other parts, including the summits of the antlers, as long as they are soft. And herein, perchance, they have stolen a march on the cooks of Paris. They get what usually goes to feed the fire. This is probably better than stall-fed beef and slaughter-house pork to make a man of. Give me a wildness whose glance no civilization can endure, — as if we lived on the marrow of koodoos devoured raw.

There are some intervals which border the strain of the wood thrush, to which I would migrate, — wild lands where no settler has squatted; to which, methinks, I am already acclimated.

The African hunter Cumming tells us that the skin of the eland, as well as that of most other antelopes just killed, emits the most delicious perfume of trees and grass. I would have every man so much like a wild antelope, so much a part and parcel of nature, that his very person should thus sweetly advertise our senses of his presence, and remind us of those parts of nature which he most haunts. I feel no disposition to be satirical, when the trapper's coat emits the odor of musquash even; it is a sweeter scent to me than that which commonly exhales from the merchant's or the scholar's garments. When I go into their wardrobes and handle their vestments, I am reminded of no grassy plains and flowery meads which they have frequented, but of dusty merchants' exchanges and libraries rather.

A tanned skin is something more than respectable, and perhaps olive is a fitter color than white for a man, — a denizen of the woods. "The pale white man!" I do not wonder that the African pitied him. Darwin the naturalist says, "A white man bathing by the side of a Tahitian was like a plant bleached by the gardener's art, compared with a fine, dark green one, growing vigorously in the open fields."

Ben Jonson exclaims, —

> "How near to good is what is fair!"

So I would say, —

> How near to good is what is *wild!*

Life consists with wildness. The most alive is the wildest. Not yet subdued to man, its presence refreshes him. One who pressed forward incessantly and never rested from his labors, who grew fast and made infinite demands on life, would always find himself in a new country or wilderness, and surrounded by the raw material of life. He would be climbing over the prostrate stems of primitive forest-trees.

Hope and the future for me are not in lawns and cultivated fields, not in towns and cities, but in the impervious and quaking swamps. When, formerly, I have analyzed my partiality for some farm which I had contemplated purchasing, I have frequently found that I was attracted solely by a few square rods of impermeable and unfathomable bog, — a natural sink in one corner of it. That was the jewel which dazzled me. I derive more of my subsistence from the swamps which surround my native town than from the cultivated gardens in the village. There are no richer parterres to my eyes than the dense beds of dwarf andromeda (*Cassandra calyculata*) which cover these tender places on the earth's surface. Botany cannot go farther than tell me the names of the shrubs which grow there, — the high blueberry, panicled andromeda, lambkill, azalea, and rhodora, — all standing in the quaking sphagnum. I often think that I should like to have my house front on this mass of dull red bushes, omitting other flower plots and borders, transplanted spruce and trim box, even graveled walks, — to have this fertile spot under my windows, not a few imported barrowfuls of soil only to cover the sand which was thrown out in digging the cellar. Why not put my house, my parlor, behind this plot, instead of behind that meagre assemblage of curiosities, that poor apology for a Nature and Art, which I call my front yard? It is an effort to clear up and make a decent appearance when the carpenter and mason have departed, though done as much for the passer-by as the dweller within. The most tasteful front-yard fence was never an agreeable object of study to me; the most elaborate ornaments, acorn tops, or what not, soon wearied and disgusted me. Bring your sills up to the very edge of the swamp, then (though it may not be the best place for a dry cellar), so that there be no access on that side to citizens. Front yards are not made to walk in, but, at most, through, and you could go in the back way.

Yes, though you may think me perverse, if it were proposed to me to dwell in the neighborhood of the most beautiful garden that ever human art contrived, or else of a Dismal Swamp, I should certainly decide for the swamp. How vain, then, have been all your labors, citizens, for me!

My spirits infallibly rise in proportion to the outward dreariness. Give me the ocean, the desert, or the wilderness! In the desert, pure air and solitude compensate for want of moisture and fertility. The traveler Burton says of it: "Your *morale* improves; you become frank and cordial, hospitable and single-minded. . . . In the desert, spirituous liquors excite only disgust. There is a keen enjoyment in a mere animal existence." They who have been traveling long on the steppes of Tartary say, "On reëntering cultivated lands, the agitation, perplexity, and turmoil of civilization oppressed and suffocated us; the air seemed to fail us, and we felt every moment as if about to die of asphyxia." When I would recreate myself, I seek the darkest wood, the thickest and most interminable and, to the citizen, most dismal, swamp. I enter a swamp as a sacred place, a *sanctum sanctorum*. There is the strength, the marrow, of Nature. The wildwood covers the virgin mould, and the same soil is good for men and for trees. A man's health requires as many acres of meadow to his prospect as his farm does loads of muck. There are the strong meats on which he feeds. A town is saved, not more by the righteous men in it than by the woods and swamps that surround it. A township where one primitive forest waves above while another primitive forest rots below,—such a town is fitted to raise not only corn and potatoes, but poets and philosophers for the coming ages. In such a soil grew Homer and Confucius and the rest, and out of such a wilderness comes the Reformer eating locusts and wild honey.

To preserve wild animals implies generally the creation of a forest for them to dwell in or resort to. So it is with man. A hundred years ago they sold bark in our streets peeled from our own woods. In the very aspect of those primitive and rugged trees there was, methinks, a tanning principle which hardened and consolidated the fibres of men's thoughts. Ah! already I shudder for these comparatively degenerate days of my native village, when you cannot collect a load of bark of good thickness, and we no longer produce tar and turpentine.

The civilized nations — Greece, Rome, England — have been sustained by the primitive forests which anciently rotted where they stand. They survive as long as the soil is not exhausted. Alas for human culture! little is to be expected of a nation, when the vegetable mould is exhausted, and it is compelled to make manure of the bones of its fathers. There the poet sustains himself merely by his own superfluous fat, and the philosopher comes down on his marrow-bones.

It is said to be the task of the American "to work

the virgin soil," and that "agriculture here already assumes proportions unknown everywhere else." I think that the farmer displaces the Indian even because he redeems the meadow, and so makes himself stronger and in some respects more natural. I was surveying for a man the other day a single straight line one hundred and thirty-two rods long, through a swamp at whose entrance might have been written the words which Dante read over the entrance to the infernal regions, "Leave all hope, ye that enter," — that is, of ever getting out again; where at one time I saw my employer actually up to his neck and swimming for his life in his property, though it was still winter. He had another similar swamp which I could not survey at all, because it was completely under water, and nevertheless, with regard to a third swamp, which I did *survey* from a distance, he remarked to me, true to his instincts, that he would not part with it for any consideration, on account of the mud which it contained. And that man intends to put a girdling ditch round the whole in the course of forty months, and so redeem it by the magic of his spade. I refer to him only as the type of a class.

The weapons with which we have gained our most important victories, which should be handed down as heirlooms from father to son, are not the sword and the lance, but the bushwhack, the turf-cutter, the spade, and the bog hoe, rusted with the blood of many a meadow, and begrimed with the dust of many a hard-fought field. The very winds blew the Indian's cornfield into the meadow, and pointed out the way which he had not the skill to follow. He had no better implement with which to intrench himself in the land than a clamshell. But the farmer is armed with plow and spade.

In literature it is only the wild that attracts us. Dullness is but another name for tameness. It is the uncivilized free and wild thinking in Hamlet and the Iliad, in all the scriptures and mythologies, not learned in the schools, that delights us. As the wild duck is more swift and beautiful than the tame, so is the wild — the mallard — thought, which 'mid falling dews wings its way above the fens. A truly good book is something as natural, and as unexpectedly and unaccountably fair and perfect, as a wild-flower discovered on the prairies of the West or in the jungles of the East. Genius is a light which makes the darkness visible, like the lightning's flash, which perchance shatters the temple of knowledge itself, — and not a taper lighted at the hearth-stone of the race, which pales before the light of common day.

English literature, from the days of the minstrels to the Lake Poets, — Chaucer and Spenser and Milton, and even Shakespeare, included, — breathes no quite fresh and, in this sense, wild strain. It is an essentially tame and civilized literature, reflecting Greece and Rome. Her wilderness is a greenwood, her wild man a Robin Hood. There is plenty of genial love of Nature, but not so much of Nature herself. Her chronicles inform us when her wild animals, but not when the wild man in her, became extinct.

The science of Humboldt is one thing, poetry is another thing. The poet to-day, notwithstanding all the discoveries of science, and the accumulated learning of mankind, enjoys no advantage over Homer.

Where is the literature which gives expression to Nature? He would be a poet who could impress the winds and streams into his service, to speak for him; who nailed words to their primitive senses, as farmers drive down stakes in the spring, which the frost has heaved; who derived his words as often as he used them,— transplanted them to his page with earth adhering to their roots; whose words were so true and fresh and natural that they would appear to expand like the buds at the approach of spring, though they lay half smothered between two musty leaves in a library, — aye, to bloom and bear fruit there, after their kind, annually, for the faithful reader, in sympathy with surrounding Nature.

I do not know of any poetry to quote which adequately expresses this yearning for the Wild. Approached from this side, the best poetry is tame. I do not know where to find in any literature, ancient or modern, any account which contents me of that Nature with which even I am acquainted. You will perceive that I demand something which no Augustan nor Elizabethan age, which no *culture*, in short, can give. Mythology comes nearer to it than anything. How much more fertile a Nature, at least, has Grecian mythology its root in than English literature! Mythology is the crop which the Old World bore before its soil was exhausted, before the fancy and imagination were affected with blight; and which it still bears, wherever its pristine vigor is unabated. All other literatures endure only as the elms which overshadow our houses; but this is like the great dragon-tree of the Western Isles, as old as mankind, and, whether that does or not, will endure as long; for the decay of other literatures makes the soil in which it thrives.

The West is preparing to add its fables to those of the East. The valleys of the Ganges, the Nile, and the Rhine having yielded their crop, it remains to be seen what the valleys of the Amazon, the Plate, the Orinoco, the St. Lawrence, and the Mississippi will produce. Perchance, when, in the course of ages, American liberty has become a fiction of the past, — as it is to some extent a fiction of the present, — the poets of the world will be inspired by American mythology.

The wildest dreams of wild men, even, are not the less true, though they may not recommend themselves to the sense which is most common among Englishmen and Americans to-day. It is not every truth that recommends itself to the common sense. Nature has a place for the wild clematis as well as for the cabbage. Some expressions of truth are reminiscent, — others merely *sensible*, as the phrase is, — others prophetic. Some forms of disease, even, may prophesy forms of health. The geologist has discovered that the figures of serpents, griffins, flying dragons, and other fanciful embellishments of heraldry, have their prototypes in the forms of fossil species which were extinct before man was created, and hence "indicate a faint and shadowy knowledge of a previous state of organic existence." The Hindoos dreamed that the earth rested on an elephant, and the elephant on a tortoise, and the tortoise on a serpent; and though it may be an unimportant coincidence, it will not be out of place here to state, that a fossil tortoise has lately been discovered in Asia large enough to support an elephant. I confess that I am partial to these wild fancies, which transcend the order of time and development. They are the sublimest recreation of the intellect. The partridge loves peas, but not those that go with her into the pot.

In short, all good things are wild and free. There is something in a strain of music, whether produced by an instrument or by the human voice, — take the sound of a bugle in a summer night, for instance, — which by its wildness, to speak without satire, reminds me of the cries emitted by wild beasts in their native forests. It is so much of their wildness as I can understand. Give me for my friends and neighbors wild men, not tame ones. The wildness of the savage is but a faint symbol of the awful ferity with which good men and lovers meet.

I love even to see the domestic animals reassert their native rights, — any evidence that they have not wholly lost their original wild habits and vigor; as when my neighbor's cow breaks out of her pasture early in the spring and boldly swims the river, a cold, gray tide, twenty-five or thirty rods wide, swollen by the melted snow. It is the buffalo crossing the Mississippi. This exploit confers some dignity on the herd in my eyes, — already dignified. The seeds of instinct are preserved under the thick hides of cattle and horses, like seeds in the bowels of the earth, an indefinite period.

Any sportiveness in cattle is unexpected. I saw one day a herd of a dozen bullocks and cows running about and frisking in unwieldy sport, like huge rats, even like kittens. They shook their heads, raised their tails, and rushed up and down a hill, and I perceived by their horns, as well as by their activity, their relation to the deer tribe. But, alas! a sudden loud *Whoa!* would have damped their ardor at once, reduced them from venison to beef, and stiffened their sides and sinews like the locomotive. Who but the Evil One has cried "Whoa!" to mankind? Indeed, the life of cattle, like that of many men, is but a sort of locomotiveness; they move a side at a time, and man, by his machinery, is meeting the horse and the ox half-way. Whatever part the whip has touched is thenceforth palsied. Who would ever think of a *side* of any of the supple cat tribe, as we speak of a *side* of beef?

I rejoice that horses and steers have to be broken before they can be made the slaves of men, and that men themselves have some wild oats still left to sow before they become submissive members of society. Undoubtedly, all men are not equally fit subjects for civilization; and because the majority, like dogs and sheep, are tame by inherited disposition, this is no reason why the others should have their natures broken that they may be reduced to the same level. Men are in the main alike, but they were made several in order that they might be various. If a low use is to be served, one man will do nearly or quite as well as another; if a high one, individual excellence is to be regarded. Any man can stop a hole to keep the wind away, but no other man could serve so rare a use as the author of this illustration did. Confucius says, "The skins of the tiger and the leopard, when they are tanned, are as the skins of the dog and the sheep tanned." But it is not the part of a true culture to tame tigers, any more than it is to make sheep ferocious; and tanning their skins for shoes is not the best use to which they can be put.

When looking over a list of men's names in a foreign language, as of military officers, or of authors who have written on a particular subject, I am reminded once more that there is nothing in a name. The name Menschikoff, for instance, has nothing in it to my ears more human than a whisker, and it may belong to a rat. As the names of the Poles and Russians are to us, so are ours to them. It is as if they had been named by the child's rigmarole, *Iery wiery ichery van, tittle-tol-tan.* I see in my mind a herd of wild creatures swarming over the earth, and to each the herdsman has affixed some barbarous sound in his own dialect. The names of men are, of course, as cheap and meaningless as *Bose* and *Tray*, the names of dogs.

Methinks it would be some advantage to philosophy if men were named merely in the gross, as they are known. It would be necessary only to know the genus and perhaps the race or variety, to know the individual. We are not prepared to believe that every private soldier in a Roman army had a name of his own, — because we have not supposed that he had a character of his own.

At present our only true names are nicknames. I knew a boy who, from his peculiar energy, was called

"Buster" by his playmates, and this rightly supplanted his Christian name. Some travelers tell us that an Indian had no name given him at first, but earned it, and his name was his fame; and among some tribes he acquired a new name with every new exploit. It is pitiful when a man bears a name for convenience merely, who has earned neither name nor fame.

I will not allow mere names to make distinctions for me, but still see men in herds for all them. A familiar name cannot make a man less strange to me. It may be given to a savage who retains in secret his own wild title earned in the woods. We have a wild savage in us, and a savage name is perchance somewhere recorded as ours. I see that my neighbor, who bears the familiar epithet William or Edwin, takes it off with his jacket. It does not adhere to him when asleep or in anger, or aroused by any passion or inspiration. I seem to hear pronounced by some of his kin at such a time his original wild name in some jaw-breaking or else melodious tongue.

Here is this vast, savage, howling mother of ours, Nature, lying all around, with such beauty, and such affection for her children, as the leopard; and yet we are so early weaned from her breast to society, to that culture which is exclusively an interaction of man on man, — a sort of breeding in and in, which produces at most a merely English nobility, a civilization destined to have a speedy limit.

In society, in the best institutions of men, it is easy to detect a certain precocity. When we should still be growing children, we are already little men. Give me a culture which imports much muck from the meadows, and deepens the soil, — not that which trusts to heating manures, and improved implements and modes of culture only!

Many a poor sore-eyed student that I have heard of would grow faster, both intellectually and physically, if, instead of sitting up so very late, he honestly slumbered a fool's allowance.

There may be an excess even of informing light. Niepce, a Frenchman, discovered "actinism," that power in the sun's rays which produces a chemical effect; that granite rocks, and stone structures, and statues of metal "are all alike destructively acted upon during the hours of sunshine, and, but for provisions of Nature no less wonderful, would soon perish under the delicate touch of the most subtile of the agencies of the universe." But he observed that "those bodies which underwent this change during the daylight possessed the power of restoring themselves to their original conditions during the hours of night, when this excitement was no longer influencing them." Hence it has been inferred that "the hours of darkness are as necessary to the inorganic creation as we know night and sleep are to the organic kingdom." Not even does the moon shine every night, but gives place to darkness.

I would not have every man nor every part of a man cultivated, any more than I would have every acre of earth cultivated: part will be tillage, but the greater part will be meadow and forest, not only serving an immediate use, but preparing a mould against a distant future, by the annual decay of the vegetation which it supports.

There are other letters for the child to learn than those which Cadmus invented. The Spaniards have a good term to express this wild and dusky knowledge, *Gramática parda*, tawny grammar, a kind of mother-wit derived from that same leopard to which I have referred.

We have heard of a Society for the Diffusion of Useful Knowledge. It is said that knowledge is power, and the like. Methinks there is equal need of a Society for the Diffusion of Useful Ignorance, what we will call Beautiful Knowledge, a knowledge useful in a higher sense: for what is most of our boasted so-called knowledge but a conceit that we know something, which robs us of the advantage of our actual ignorance? What we call knowledge is often our positive ignorance; ignorance our negative knowledge. By long years of patient industry and reading of the newspapers, — for what are the libraries of science but files of newspapers? — a man accumulates a myriad facts, lays them up in his memory, and then when in some spring of his life he saunters abroad into the Great Fields of thought, he, as it were, goes to grass like a horse and leaves all his harness behind in the stable. I would say to the Society for the Diffusion of Useful Knowledge, sometimes, — Go to grass. You have eaten hay long enough. The spring has come with its green crop. The very cows are driven to their country pastures before the end of May; though I have heard of one unnatural farmer who kept his cow in the barn and fed her on hay all the year round. So, frequently, the Society for the Diffusion of Useful Knowledge treats its cattle.

A man's ignorance sometimes is not only useful, but beautiful, — while his knowledge, so called, is oftentimes worse than useless, besides being ugly. Which is the best man to deal with, — he who knows nothing about a subject, and, what is extremely rare, knows that he knows nothing, or he who really knows something about it, but thinks that he knows all?

My desire for knowledge is intermittent, but my desire to bathe my head in atmospheres unknown to my feet is perennial and constant. The highest that we can attain to is not Knowledge, but Sympathy with Intelligence. I do not know that this higher knowledge amounts to anything more definite than a novel and grand surprise on a sudden revelation of the insufficiency of all that we called Knowledge before, — a discovery that there are more things in heaven and earth than are dreamed of in our philosophy. It is the lighting up of the mist by the sun. Man cannot *know* in any higher sense than this, any more than he can look serenely and with impunity in the face of the sun: Ὡς τὶ νοῶν, οὐ κεῖνον νοήσεις, "You will not perceive that, as perceiving a particular thing," say the Chaldean Oracles.

There is something servile in the habit of seeking after a law which we may obey. We may study the laws of matter at and for our convenience, but a successful life knows no law. It is an unfortunate discovery certainly, that of a law which binds us where we did not know before that we were bound. Live free, child of the mist, — and with respect to knowledge we are all children of the mist. The man who takes the liberty to live is superior to all the laws, by virtue of his relation to the lawmaker. "That is active duty," says the Vishnu Purana, "which is not for our bondage; that is knowledge which is for our liberation: all other duty is good only unto weariness; all other knowledge is only the cleverness of an artist."

It is remarkable how few events or crises there are in our histories, how little exercised we have been in our minds, how few experiences we have had. I would fain be assured that I am growing apace and rankly, though my very growth disturb this dull equanimity, — though it be with struggle through long, dark, muggy nights or seasons of gloom. It would be well if all our lives were a divine tragedy even, instead of this trivial comedy or farce. Dante, Bunyan, and others appear to have been exercised in their minds more than we: they were subjected to a kind of culture such as our district schools and colleges do not contemplate. Even Mahomet, though many may scream at his name, had a good deal more to live for, aye, and to die for, than they have commonly.

When, at rare intervals, some thought visits one, as perchance he is walking on a railroad, then, indeed, the cars go by without his hearing them. But soon, by some inexorable law, our life goes by and the cars return.

> "Gentle breeze, that wanderest unseen,
> And bendest the thistles round Loira of storms,
> Traveler of the windy glens,
> Why hast thou left my ear so soon?"

While almost all men feel an attraction drawing them to society, few are attracted strongly to Nature. In their reaction to Nature men appear to me for the most part, notwithstanding their arts, lower than the animals. It is not often a beautiful relation, as in the case of the animals. How little appreciation of the beauty of the landscape there is among us! We have to be told that the Greeks called the world Κόσμος, Beauty, or Order, but we do not see clearly why they did so, and we esteem it at best only a curious philological fact.

For my part, I feel that with regard to Nature I live a sort of border life, on the confines of a world into which I make occasional and transient forays only, and my patriotism and allegiance to the state into whose territories I seem to retreat are those of a moss-trooper. Unto a life which I call natural I would gladly follow even a will-o'-the-wisp through bogs and sloughs unimaginable, but no moon nor firefly has shown me the causeway to it. Nature is a personality so vast and universal that we have never seen one of her features. The walker in the familiar fields which stretch around my native town sometimes finds himself in another land than is described in their owners' deeds, as it were in some faraway field on the confines of the actual Concord, where her jurisdiction ceases, and the idea which the word Concord suggests ceases to be suggested. These farms which I have myself surveyed, these bounds which I have set up, appear dimly still as through a mist; but they have no chemistry to fix them; they fade from the surface of the glass, and the picture which the painter painted stands out dimly from beneath. The world with which we are commonly acquainted leaves no trace, and it will have no anniversary.

I took a walk on Spaulding's Farm the other afternoon. I saw the setting sun lighting up the opposite side of a stately pine wood. Its golden rays straggled into the aisles of the wood as into some noble hall. I was impressed as if some ancient and altogether admirable and shining family had settled there in that part of the land called Concord, unknown to me, — to whom the sun was servant, — who had not gone into society in the village, — who had not been called on. I saw their park, their pleasure-ground, beyond through the wood, in Spaulding's cranberry-meadow. The pines furnished them with gables as they grew. Their house was not obvious to vision; the trees grew through it. I do not know whether I heard the sounds of a suppressed hilarity or not. They seemed to recline on the sunbeams. They have sons and daughters. They are quite well. The farmer's cart-path, which leads directly through their hall, does not in the least put them out, as the muddy bottom of a pool is sometimes seen through the reflected skies. They never heard of Spaulding, and do not know that he is their neighbor, — notwithstanding I heard him whistle as he drove his team through the house. Nothing can equal the serenity of their lives. Their coat-of-arms is simply a lichen. I saw it painted on the pines and oaks. Their attics were in the tops of the trees. They are of no politics. There was no noise of labor. I did not perceive that they were weaving or spinning. Yet I did detect, when the wind lulled and hearing was done away, the finest imaginable sweet musical hum, — as of a distant hive in May, — which perchance was the sound of their thinking. They had no idle thoughts, and no one without could see their work, for their industry was not as in knots and excrescences embayed.

But I find it difficult to remember them. They fade irrevocably out of my mind even now while I speak, and endeavor to recall them and recollect myself. It is only after a long and serious effort to recollect my best thoughts that I become again aware of their cohabitancy. If it were not for such families as this, I think I should move out of Concord.

We are accustomed to say in New England that few and fewer pigeons visit us every year. Our forests furnish no mast for them. So, it would seem, few and fewer thoughts visit each growing man from year to year, for the grove in our minds is laid waste, — sold to feed unnecessary fires of ambition, or sent to mill, — and there is scarcely a twig left for them to perch on. They no longer build nor breed with us. In some more genial season, perchance, a faint shadow flits across the landscape of the mind, cast by the *wings* of some thought in its vernal or autumnal migration, but, looking up, we are unable to detect the substance of the thought itself. Our winged thoughts are turned to poultry. They no longer soar, and they attain only to a Shanghai and Cochin-China grandeur. Those *gra-a-ate thoughts*, those *gra-a-ate men* you hear of!

We hug the earth, — how rarely we mount! Methinks we might elevate ourselves a little more. We might climb a tree, at least. I found my account in climbing a tree once. It was a tall white pine, on the top of a hill; and though I got well pitched, I was well paid for it, for I discovered new mountains in the horizon which I had never seen before, — so much more of the earth and the heavens. I might have walked about the foot of the tree for threescore years and ten, and yet I certainly should never have seen them. But, above all, I discovered around me, — it was near the end of June, — on the ends of the topmost branches only, a few minute and delicate red cone-like blossoms, the fertile flower of the white pine looking heavenward. I carried straightway to the village the topmost spire, and showed it to stranger jurymen who walked the streets, — for it was court week, — and to farmers and lumber-dealers and woodchoppers and hunters, and not one had ever seen the like before, but they wondered as at a star dropped down. Tell of ancient architects finishing their works on the tops of columns as perfectly as on the lower and more visible parts! Nature has from the first expanded the minute blossoms of the forest only toward the heavens, above men's heads and unobserved by them. We see only the flowers that are under our feet in the meadows. The pines have developed their delicate blossoms on the highest twigs of the wood every summer for ages, as well over the heads of Nature's red children as of her white ones; yet scarcely a farmer or hunter in the land has ever seen them.

Above all, we cannot afford not to live in the present. He is blessed over all mortals who loses no moment of the passing life in remembering the past. Unless our philosophy hears the cock crow in every barn-yard within our horizon, it is belated. That sound commonly reminds us that we are growing rusty and antique in our employments and habits of thought. His philosophy comes down to a more recent time than ours. There is something suggested by it that is a newer testament, — the gospel according to this moment. He has not fallen astern; he has got up early and kept up early, and to be where he is is to be in season, in the foremost rank of time. It is an expression of the health and soundness of Nature, a brag for all the world, — healthiness as of a spring burst forth, a new fountain of the Muses, to celebrate this last instant of time. Where he lives no fugitive slave laws are passed. Who has not betrayed his master many times since last he heard that note?

The merit of this bird's strain is in its freedom from all plaintiveness. The singer can easily move us to tears or to laughter, but where is he who can excite in us a pure morning joy? When, in doleful dumps, breaking the awful stillness of our wooden sidewalk on a Sunday, or, perchance, a watcher in the house of mourning, I hear a cockerel crow far or near, I think to myself, "There is one of us well, at any rate," — and with a sudden gush return to my senses.

We had a remarkable sunset one day last November. I was walking in a meadow, the source of a small brook, when the sun at last, just before setting, after a cold, gray day, reached a clear stratum in the horizon, and the softest, brightest morning sunlight fell on the dry grass and on the stems of the trees in the opposite horizon and on the leaves of the shrub oaks on the hillside, while our shadows stretched long over the meadow eastward, as if we were the only motes in its beams. It was such a light as we could not have imagined a moment before, and the air also was so warm and serene that nothing was wanting to make a paradise of that meadow. When we reflected that this was not a solitary phenomenon, never to happen again, but that it would happen forever and ever, an infinite number of evenings, and cheer and reassure the latest child that walked there, it was more glorious still.

The sun sets on some retired meadow, where no house is visible, with all the glory and splendor that it lavishes on cities, and perchance as it has never set before, — where there is but a solitary marsh hawk to have his wings gilded by it, or only a musquash looks out from his cabin, and there is some little black-veined brook in the midst of the marsh, just beginning to meander, winding slowly round a decaying stump. We walked in so pure and bright a light, gilding the withered grass and leaves, so softly and serenely bright, I thought I had never bathed in such a golden flood, without a ripple or a murmur to it. The west side of every wood and rising ground gleamed like the boundary of Elysium, and the sun on our backs seemed like a gentle herdsman driving us home at evening.

So we saunter toward the Holy Land, till one day the sun shall shine more brightly than ever he has done, shall perchance shine into our minds and hearts, and light up our whole lives with a great awakening light, as warm and serene and golden as on a bankside in autumn.

AUTUMNAL TINTS

EUROPEANS coming to America are surprised by the brilliancy of our autumnal foliage. There is no account of such a phenomenon in English poetry, because the trees acquire but few bright colors there. The most that Thomson says on this subject in his "Autumn" is contained in the lines, —

> "But see the fading many-colored woods
> Shade deepening over shade, the country round
> Imbrown; a crowded umbrage, dusk and dun,
> Of every hue, from wan declining green
> To sooty dark;"

and in the line in which he speaks of

> "Autumn beaming o'er the yellow woods."

The autumnal change of our woods has not made a deep impression on our own literature yet. October has hardly tinged our poetry.

A great many, who have spent their lives in cities, and have never chanced to come into the country at this season, have never seen this, the flower, or rather the ripe fruit, of the year. I remember riding with one such citizen, who, though a fortnight too late for the most brilliant tints, was taken by surprise, and would not believe that there had been any brighter. He had never heard of this phenomenon before. Not only many in our towns have never witnessed it, but it is scarcely remembered by the majority from year to year.

Most appear to confound changed leaves with withered ones, as if they were to confound ripe apples with rotten ones. I think that the change to some higher color in a leaf is an evidence that it has arrived at a late and perfect maturity, answering to the maturity of fruits. It is generally the lowest and oldest leaves which change first. But as the perfect-winged and usually bright-colored insect is short-lived, so the leaves ripen but to fall.

Generally, every fruit, on ripening, and just before it falls, when it commences a more independent and individual existence, requiring less nourishment from any source, and that not so much from the earth through its stem as from the sun and air, acquires a bright tint. So do leaves. The physiologist says it is "due to an increased absorption of oxygen." That is the scientific account of the matter, — only a reassertion of the fact. But I am more interested in the rosy cheek than I am to know what particular diet the maiden fed on. The very forest and herbage, the pellicle of the earth, must acquire a bright color, an evidence of its ripeness, — as if the globe itself were a fruit on its stem, with ever a cheek toward the sun.

Flowers are but colored leaves, fruits but ripe ones. The edible part of most fruits is, as the physiologist says, "the parenchyma or fleshy tissue of the leaf," of which they are formed.

Our appetites have commonly confined our views of ripeness and its phenomena, color, mellowness, and perfectness, to the fruits which we eat, and we are wont to forget that an immense harvest which we do not eat, hardly use at all, is annually ripened by Nature. At our annual cattle-shows and horticultural exhibitions, we make, as we think, a great show of fair fruits, destined, however, to a rather ignoble end, fruits not valued for their beauty chiefly. But round about and within our towns there is annually another show of fruits, on an infinitely grander scale, fruits which address our taste for beauty alone.

October is the month for painted leaves. Their rich glow now flashes round the world. As fruits and leaves and the day itself acquire a bright tint just before they fall, so the year near its setting. October is its sunset sky; November the later twilight.

I formerly thought that it would be worth the while to get a specimen leaf from each changing tree, shrub, and herbaceous plant, when it had acquired its brightest characteristic color, in its transition from the green to the brown state, outline it, and copy its color exactly, with paint, in a book, which should be entitled "October, or Autumnal Tints," — beginning with the earliest reddening woodbine and the lake of radical leaves, and coming down through the maples, hickories, and sumachs, and many beautifully freckled leaves less generally known, to the latest oaks and aspens. What a memento such a book would be! You would need only to turn over its leaves to take a ramble through the autumn woods whenever you pleased. Or if I could preserve the leaves themselves, unfaded, it would be better still. I have made but little progress toward such a book, but I have endeavored, instead, to describe all these bright tints in the order in which they present themselves. The following are some extracts from my notes.

THE PURPLE GRASSES

By the twentieth of August, everywhere in woods and swamps we are reminded of the fall, both by the richly spotted sarsaparilla leaves and brakes, and the withering and blackened skunk-cabbage and hellebore, and, by the riverside, the already blackening pontederia.

The purple grass (*Eragrostis pectinacea*) is now in the height of its beauty. I remember still when I first noticed this grass particularly. Standing on a hillside near our river, I saw, thirty or forty rods off, a stripe of purple half a dozen rods long, under the edge of a wood, where the ground sloped toward a meadow. It was as high-colored and interesting, though not quite so bright, as the patches of rhexia, being a darker purple, like a berry's stain laid on close and thick. On going to and examining it, I found it to be a kind of grass in bloom, hardly a foot high, with but few green blades, and a fine spreading panicle of purple flowers, a shallow, purplish mist trembling around me. Close at hand it appeared but a dull purple, and made little impression on the eye; it was even difficult to detect; and if you plucked a single plant, you were surprised to find how thin it was, and how little color it had. But viewed at a distance in a favorable light, it was of a fine lively purple, flower-like, enriching the earth. Such puny causes combine to produce these decided effects. I was the more surprised and charmed because grass is commonly of a sober and humble color.

With its beautiful purple blush it reminds me, and

supplies the place, of the rhexia, which is now leaving off, and it is one of the most interesting phenomena of August. The finest patches of it grow on waste strips or selvages of land at the base of dry hills, just above the edge of the meadows, where the greedy mower does not deign to swing his scythe; for this is a thin and poor grass, beneath his notice. Or, it may be, because it is so beautiful he does not know that it exists; for the same eye does not see this and timothy. He carefully gets the meadow-hay and the more nutritious grasses which grow next to that, but he leaves this fine purple mist for the walker's harvest,— fodder for his fancy stock. Higher up the hill, perchance, grow also blackberries, John's-wort, and neglected, withered, and wiry June-grass. How fortunate that it grows in such places, and not in the midst of the rank grasses which are annually cut! Nature thus keeps use and beauty distinct. I know many such localities, where it does not fail to present itself annually, and paint the earth with its blush. It grows on the gentle slopes, either in a continuous patch or in scattered and rounded tufts a foot in diameter, and it lasts till it is killed by the first smart frosts.

In most plants the corolla or calyx is the part which attains the highest color, and is the most attractive; in many it is the seed-vessel or fruit; in others, as the red maple, the leaves; and in others still it is the very culm itself which is the principal flower or blooming part.

The last is especially the case with the poke or garget (*Phytolacca decandra*). Some which stand under our cliffs quite dazzle me with their purple stems now and early in September. They are as interesting to me as most flowers, and one of the most important fruits of our autumn. Every part is flower (or fruit), such is its superfluity of color, — stem, branch, peduncle, pedicel, petiole, and even the at length yellowish, purple-veined leaves. Its cylindrical racemes of berries of various hues, from green to dark purple, six or seven inches long, are gracefully drooping on all sides, offering repasts to the birds; and even the sepals from which the birds have picked the berries are a brilliant lake red, with crimson flame-like reflections, equal to anything of the kind, — all on fire with ripeness. Hence the *lacca*, from *lac*, lake. There are at the same time flower-buds, flowers, green berries, dark-purple or ripe ones, and these flower-like sepals, all on the same plant.

We love to see any redness in the vegetation of the temperate zone. It is the color of colors. This plant speaks to our blood. It asks a bright sun on it to make it show to best advantage, and it must be seen at this season of the year. On warm hillsides its stems are ripe by the twenty-third of August. At that date I walked through a beautiful grove of them, six or seven feet high, on the side of one of our cliffs, where they ripen early. Quite to the ground they were a deep, brilliant purple, with a bloom contrasting with the still clear green leaves. It appears a rare triumph of Nature to have produced and perfected such a plant, as if this were enough for a summer. What a perfect maturity it arrives at! It is the emblem of a successful life concluded by a death not premature, which is an ornament to Nature. What if we were to mature as perfectly, root and branch, glowing in the midst of our decay, like the poke! I confess that it excites me to behold them. I cut one for a cane, for I would fain handle and lean on it. I love to press the berries between my fingers, and see their juice staining my hand. To walk amid these upright, branching casks of purple wine, which retain and diffuse a sunset glow, tasting each one with your eye, instead of counting the pipes on a London dock, what a privilege! For Nature's vintage is not confined to the vine. Our poets have sung of wine, the product of a foreign plant which commonly they never saw, as if our own plants had no juice in them more than the singers. Indeed, this has been called by some the American grape, and, though a native of America, its juices are used in some foreign countries to improve the color of the wine; so that the poetaster may be celebrating the virtues of the poke without knowing it. Here are berries enough to paint afresh the western sky, and play the bacchanal with, if you will. And what flutes its ensanguined stems would make, to be used in such a dance! It is truly a royal plant. I could spend the evening of the year musing amid the poke stems. And perchance amid these groves might arise at last a new school of philosophy or poetry. It lasts all through September.

At the same time with this, or near the end of August, a to me very interesting genus of grasses, andropogons, or beard-grasses, is in its prime: *Andropogon furcatus*, forked beard-grass, or call it purple-fingered grass; *Andropogon scoparius*, purple wood-grass; and *Andropogon* (now called *Sorghum*) *nutans*, Indian-grass.

The first is a very tall and slender-culmed grass, three to seven feet high, with four or five purple finger-like spikes raying upward from the top. The second is also quite slender, growing in tufts two feet high by one wide, with culms often somewhat curving, which, as the spikes go out of bloom, have a whitish, fuzzy look. These two are prevailing grasses at this season on dry and sandy fields and hillsides. The culms of both, not to mention their pretty flowers, reflect a purple tinge, and help to declare the ripeness of the year. Perhaps I have the more sympathy with them because they are despised by the farmer, and occupy sterile and neglected soil. They are high-colored, like ripe grapes, and express a maturity which the spring did not suggest. Only the August sun could have thus burnished these culms and leaves. The farmer has long since done his upland haying, and he will not condescend to bring his scythe to where these slender wild grasses have at length flowered thinly; you often see spaces of bare sand amid them. But I walk encouraged between the tufts of purple wood-grass over the sandy fields, and along the edge of the shrub oaks, glad to recognize these simple contemporaries. With thoughts cutting a broad swathe I "get" them, with horse-raking thoughts I gather them into windrows. The fine-eared poet may hear the whetting of my scythe. These two were almost the first grasses that I learned to distinguish, for I had not known by how many friends I was surrounded; I had seen them simply as grasses standing. The purple of their culms also excites me like that of the poke-weed stems.

Think what refuge there is for one, before August is over, from college commencements and society that isolates! I can skulk amid the tufts of purple wood-grass on the borders of the "Great Fields." Wherever I walk these afternoons, the purple-fingered grass also stands like a guide-board, and points my thoughts to more poetic paths than they have lately traveled.

A man shall perhaps rush by and trample down plants as high as his head, and cannot be said to know that they exist, though he may have cut many tons of them, littered his stables with them, and fed them to his cattle for years. Yet, if he ever favorably attends to them, he may be overcome by their beauty. Each humblest plant, or weed, as we call it, stands there to express some thought or mood of ours; and yet how long it stands in vain! I had walked over those Great Fields so many Augusts, and never yet distinctly recognized these purple companions that I had there. I had brushed against them and trodden on them, forsooth; and now, at last, they, as it were, rose up and blessed me. Beauty and true wealth are always thus cheap and despised. Heaven might be defined as the place which men avoid. Who can doubt that these grasses, which the farmer says are of no account to him, find some compensation in your appreciation of them? I may say that I never saw them before; though, when I came to look them face to face, there did come down to me a purple gleam from previous years; and now, wherever I go, I see hardly anything else. It is the reign and presidency of the andropogons.

Almost the very sands confess the ripening influence

of the August sun, and methinks, together with the slender grasses waving over them, reflect a purple tinge. The impurpled sands! Such is the consequence of all this sunshine absorbed into the pores of plants and of the earth. All sap or blood is now wine-colored. At last we have not only the purple sea, but the purple land.

The chestnut beard-grass, Indian-grass, or wood-grass, growing here and there in waste places, but more rare than the former (from two to four or five feet high), is still handsomer and of more vivid colors than its congeners, and might well have caught the Indian's eye. It has a long, narrow, one-sided, and slightly nodding panicle of bright purple and yellow flowers, like a banner raised above its reedy leaves. These bright standards are now advanced on the distant hillsides, not in large armies, but in scattered troops or single file, like the red men. They stand thus fair and bright, representative of the race which they are named after, but for the most part unobserved as they. The expression of this grass haunted me for a week, after I first passed and noticed it, like the glance of an eye. It stands like an Indian chief taking a last look at his favorite hunting-grounds.

THE RED MAPLE

By the twenty-fifth of September, the red maples generally are beginning to be ripe. Some large ones have been conspicuously changing for a week, and some single trees are now very brilliant. I notice a small one, half a mile off across a meadow, against the green woodside there, a far brighter red than the blossoms of any tree in summer, and more conspicuous. I have observed this tree for several autumns invariably changing earlier than its fellows, just as one tree ripens its fruit earlier than another. It might serve to mark the season, perhaps. I should be sorry if it were cut down. I know of two or three such trees in different parts of our town, which might, perhaps, be propagated from, as early ripeners or September trees, and their seed be advertised in the market, as well as that of radishes, if we cared as much about them.

At present these burning bushes stand chiefly along the edge of the meadows, or I distinguish them afar on the hillsides here and there. Sometimes you will see many small ones in a swamp turned quite crimson when all other trees around are still perfectly green, and the former appear so much the brighter for it. They take you by surprise, as you are going by on one side, across the fields, thus early in the season, as if it were some gay encampment of the red men, or other foresters, of whose arrival you had not heard.

Some single trees, wholly bright scarlet, seen against others of their kind still freshly green, or against evergreens, are more memorable than whole groves will be by and by. How beautiful, when a whole tree is like one great scarlet fruit full of ripe juices, every leaf, from lowest limb to topmost spire, all aglow, especially if you look toward the sun! What more remarkable object can there be in the landscape? Visible for miles, too fair to be believed. If such a phenomenon occurred but once, it would be handed down by tradition to posterity, and get into the mythology at last.

The whole tree thus ripening in advance of its fellows attains a singular preëminence, and sometimes maintains it for a week or two. I am thrilled at the sight of it, bearing aloft its scarlet standard for the regiment of green-clad foresters around, and I go half a mile out of my way to examine it. A single tree becomes thus the crowning beauty of some meadowy vale, and the expression of the whole surrounding forest is at once more spirited for it.

A small red maple has grown, perchance, far away at the head of some retired valley, a mile from any road, unobserved. It has faithfully discharged the duties of a maple there, all winter and summer, neglected none of its economies, but added to its stature in the virtue which belongs to a maple, by a steady growth for so many months, never having gone gadding abroad, and is nearer heaven than it was in the spring. It has faithfully husbanded its sap, and afforded a shelter to the wandering bird, has long since ripened its seeds and committed them to the winds, and has the satisfaction of knowing, perhaps, that a thousand little well-behaved maples are already settled in life somewhere. It deserves well of Mapledom. Its leaves have been asking it from time to time, in a whisper, "When shall we redden?" And now, in this month of September, this month of traveling, when men are hastening to the seaside, or the mountains, or the lakes, this modest maple, still without budging an inch, travels in its reputation, — runs up its scarlet flag on that hillside, which shows that it has finished its summer's work before all other trees, and withdraws from the contest. At the eleventh hour of the year, the tree which no scrutiny could have detected here when it was most industrious is thus, by the tint of its maturity, by its very blushes, revealed at last to the careless and distant traveler, and leads his thoughts away from the dusty road into those brave solitudes which it inhabits. It flashes out conspicuous with all the virtue and beauty of a maple,—*Acer rubrum.* We may now read its title, or *rubric*, clear. Its *virtues*, not its sins, are as scarlet.

Notwithstanding the red maple is the most intense scarlet of any of our trees, the sugar maple has been the most celebrated, and Michaux in his "Sylva" does not speak of the autumnal color of the former. About the second of October, these trees, both large and small, are most brilliant, though many are still green. In "sprout-lands" they seem to vie with one another, and ever some particular one in the midst of the crowd will be of a peculiarly pure scarlet, and by its more intense color attract our eye even at a distance, and carry off the palm. A large red maple swamp, when at the height of its change, is the most obviously brilliant of all tangible things, where I dwell, so abundant is this tree with us. It varies much both in form and color. A great many are merely yellow; more, scarlet; others, scarlet deepening into crimson, more red than common. Look at yonder swamp of maples mixed with pines, at the base of a pine-clad hill, a quarter of a mile off, so that you get the full effect of the bright colors, without detecting the imperfections of the leaves, and see their yellow, scarlet, and crimson fires, of all tints, mingled and contrasted with the green. Some maples are yet green, only yellow or crimson-tipped on the edges of their flakes, like the edges of a hazel-nut bur; some are wholly brilliant scarlet, raying out regularly and finely every way, bilaterally, like the veins of a leaf; others, of more irregular form, when I turn my head slightly, emptying out some of its earthiness and concealing the trunk of the tree, seem to rest heavily flake on flake, like yellow and scarlet clouds, wreath upon wreath, or like snow-drifts driving through the air, stratified by the wind. It adds greatly to the beauty of such a swamp at this season, that, even though there may be no other trees interspersed, it is not seen as a simple mass of color, but, different trees being of different colors and hues, the outline of each crescent treetop is distinct, and where one laps on to another. Yet a painter would hardly venture to make them thus distinct a quarter of a mile off.

As I go across a meadow directly toward a low rising ground this bright afternoon, I see, some fifty rods off toward the sun, the top of a maple swamp just appearing over the sheeny russet edge of the hill, a stripe apparently twenty rods long by ten feet deep, of the most intensely brilliant scarlet, orange, and yellow, equal to any flowers or fruits, or any tints ever painted. As I advance, lowering the edge of the hill which makes the firm foreground or lower frame of the picture, the depth of the brilliant grove revealed steadily increases, suggesting that the whole of the inclosed valley is filled with such color. One wonders that the tithing-men and fathers of the town are not out to see what the trees mean by their high colors and exuberance of spirits, fearing that some mischief is brewing. I do not see what the Puritans did at this season, when the maples blaze out in scarlet. They certainly could not have worshiped in groves then. Perhaps that is what they built meeting-houses and fenced them round with horse-sheds for.

THE ELM

Now too, the first of October, or later, the elms are at the height of their autumnal beauty, — great brownish-yellow masses, warm from their September oven, hanging over the highway. Their leaves are perfectly ripe. I wonder if there is any answering ripeness in the lives of the men who live beneath them. As I look down our street, which is lined with them, they remind me both by their form and color of yellowing sheaves of grain, as if the harvest had indeed come to the village itself, and we might expect to find some maturity and *flavor* in the thoughts of the villagers at last. Under those bright rustling yellow piles just ready to fall on the heads of the walkers, how can any crudity or greenness of thought or act prevail? When I stand where half a dozen large elms droop over a house, it is as if I stood within a ripe pumpkin-rind, and I feel as mellow as if I were the pulp, though I may be somewhat stringy and seedy withal. What is the late greenness of the English elm, like a cucumber out of season, which does not know when to have done, compared with the early and golden maturity of the American tree? The street is the scene of a great harvest-home. It would be worth the while to set out these trees, if only for their autumnal value. Think of these great yellow canopies or parasols held over our heads and houses by the mile together, making the village all one and compact, — an *ulmarium*, which is at the same time a nursery of men! And then how gently and unobserved they drop their burden and let in the sun when it is wanted, their leaves not heard when they fall on our roofs and in our streets; and thus the village parasol is shut up and put away! I see the market-man driving into the village, and disappearing under its canopy of elm-tops, with *his* crop, as into a great granary or barn-yard. I am tempted to go thither as to a husking of thoughts, now dry and ripe, and ready to be separated from their integuments; but, alas! I foresee that it will be chiefly husks and little thought, blasted pig-corn, fit only for cob-meal, — for, as you sow, so shall you reap.

FALLEN LEAVES

By the sixth of October the leaves generally begin to fall, in successive showers, after frost or rain; but the principal leaf-harvest, the acme of the *Fall*, is commonly about the sixteenth. Some morning at that date there is perhaps a harder frost than we have seen, and ice formed under the pump, and now, when the morning wind rises, the leaves come down in denser showers than ever. They suddenly form thick beds or carpets on the ground, in this gentle air, or even without wind, just the size and form of the tree above. Some trees, as small hickories, appear to have dropped their leaves instantaneously, as a soldier grounds arms at a signal;

and those of the hickory, being bright yellow still, though withered, reflect a blaze of light from the ground where they lie. Down they have come on all sides, at the first earnest touch of autumn's wand, making a sound like rain.

Or else it is after moist and rainy weather that we notice how great a fall of leaves there has been in the night, though it may not yet be the touch that loosens the rock maple leaf. The streets are thickly strewn with the trophies, and fallen elm leaves make a dark brown pavement under our feet. After some remarkably warm Indian-summer day or days, I perceive that it is the unusual heat which, more than anything, causes the leaves to fall, there having been, perhaps, no frost nor rain for some time. The intense heat suddenly ripens and wilts them, just as it softens and ripens peaches and other fruits, and causes them to drop.

The leaves of late red maples, still bright, strew the earth, often crimson-spotted on a yellow ground, like some wild apples, — though they preserve these bright colors on the ground but a day or two, especially if it rains. On causeways I go by trees here and there all bare and smoke-like, having lost their brilliant clothing; but there it lies, nearly as bright as ever, on the ground on one side, and making nearly as regular a figure as lately on the tree. I would rather say that I first observe the trees thus flat on the ground like a permanent colored shadow, and they suggest to look for the boughs that bore them. A queen might be proud to walk where these gallant trees have spread their bright cloaks in the mud. I see wagons roll over them as a shadow or a reflection, and the drivers heed them just as little as they did their shadows before.

Birds' nests, in the huckleberry and other shrubs, and in trees, are already being filled with the withered leaves. So many have fallen in the woods that a squirrel cannot run after a falling nut without being heard. Boys are raking them in the streets, if only for the pleasure of dealing with such clean, crisp substances. Some sweep the paths scrupulously neat, and then stand to see the next breath strew them with new trophies. The swamp floor is thickly covered, and the *Lycopodium lucidulum* looks suddenly greener amid them. In dense woods they half cover pools that are three or four rods long. The other day I could hardly find a well-known spring, and even suspected that it had dried up, for it was completely concealed by freshly fallen leaves; and when I swept them aside and revealed it, it was like striking the earth, with Aaron's rod, for a new spring. Wet grounds about the edges of swamps look dry with them. At one swamp, where I was surveying, thinking to step on a leafy shore from a rail, I got into the water more than a foot deep.

When I go to the river the day after the principal fall of leaves, the sixteenth, I find my boat all covered, bottom and seats, with the leaves of the golden willow under which it is moored, and I set sail with a cargo of them rustling under my feet. If I empty it, it will be full again to-morrow. I do not regard them as litter, to be swept out, but accept them as suitable straw or matting for the bottom of my carriage. When I turn up into the mouth of the Assabet, which is wooded, large fleets of leaves are floating on its surface, as it were getting out to sea, with room to tack; but next the shore, a little farther up, they are thicker than foam, quite concealing the water for a rod in width, under and amid the alders, button-bushes, and maples, still perfectly light and dry, with fibre unrelaxed; and at a rocky bend where they are met and stopped by the morning wind, they sometimes form a broad and dense crescent quite across the river. When I turn my prow that way, and the wave which it makes strikes them, list what a pleasant rustling from these dry substances getting on one another! Often it is their undulation only which reveals the water beneath them. Also every motion of the wood turtle on the shore is betrayed by their rustling there. Or even in mid-channel, when the wind rises, I hear them blown with a rustling sound. Higher up they are slowly moving round and round in some great eddy which the river makes, as that at the "Leaning Hemlocks," where the water is deep, and the current is wearing into the bank.

Perchance, in the afternoon of such a day, when the water is perfectly calm and full of reflections, I paddle gently down the main stream, and, turning up the Assabet, reach a quiet cove, where I unexpectedly find myself surrounded by myriads of leaves, like fellow-voyagers, which seem to have the same purpose, or want of purpose, with myself. See this great fleet of scattered leaf-boats which we paddle amid, in this smooth river-bay, each one curled up on every side by the sun's skill, each nerve a stiff spruce knee, — like boats of hide, and of all patterns, — Charon's boat probably among the rest, — and some with lofty prows and poops, like the stately vessels of the ancients, scarcely moving in the sluggish current, — like the great fleets, the dense Chinese cities of boats, with which you mingle on entering some great mart, some New York or Canton, which we are all steadily approaching together. How gently each has been deposited on the water! No violence has been used towards them yet, though, perchance, palpitating hearts were present at the launching. And painted ducks, too, the splendid wood duck among the rest, often come to sail and float amid the painted leaves, — barks of a nobler model still!

What wholesome herb drinks are to be had in the swamps now! What strong medicinal but rich scents from the decaying leaves! The rain falling on the freshly dried herbs and leaves, and filling the pools and ditches into which they have dropped thus clean and rigid, will soon convert them into tea, — green, black, brown, and yellow teas, of all degrees of strength, enough to set all Nature a-gossiping. Whether we drink them or not, as yet, before their strength is drawn, these leaves, dried on great Nature's coppers, are of such various pure and delicate tints as might make the fame of Oriental teas.

How they are mixed up, of all species, oak and maple and chestnut and birch! But Nature is not cluttered with them; she is a perfect husbandman; she stores them all. Consider what a vast crop is thus annually shed on the earth! This, more than any mere grain or seed, is the great harvest of the year. The trees are now repaying the earth with interest what they have taken from it. They are discounting. They are about to add a leaf's thickness to the depth of the soil. This is the beautiful way in which Nature gets her muck, while I chaffer with this man and that, who talks to me about sulphur and the cost of carting. We are all the richer for their decay. I am more interested in this crop than in the English grass alone or in the corn. It prepares the virgin mould for future corn-fields and forests, on which the earth fattens. It keeps our homestead in good heart.

For beautiful variety no crop can be compared with this. Here is not merely the plain yellow of the grains, but nearly all the colors that we know, the brightest blue not excepted: the early blushing maple, the poison sumach blazing its sins as scarlet, the mulberry ash, the rich chrome yellow of the poplars, the brilliant red huckleberry, with which the hills' backs are painted, like those of sheep. The frost touches them, and, with the slightest breath of returning day or jarring of earth's axle, see in what showers they come floating down! The ground is all parti-colored with them. But they still live in the soil, whose fertility and bulk they increase, and in the forests that spring from it. They stoop to rise, to mount higher in coming years, by subtle chemistry, climbing by the sap in the trees; and the sapling's first fruits thus shed, transmuted at last, may adorn its crown, when, in after years, it has become the monarch of the forest.

It is pleasant to walk over the beds of these fresh, crisp, and rustling leaves. How beautifully they go to their graves! how gently lay themselves down and turn to mould! — painted of a thousand hues, and fit to make the beds of us living. So they troop to their last resting-place, light and frisky. They put on no weeds, but merrily they go scampering over the earth, selecting the spot, choosing a lot, ordering no iron fence, whispering all through the woods about it, — some choosing the spot where the bodies of men are mouldering beneath, and meeting them half-way. How many flutterings before they rest quietly in their graves! They that soared so loftily, how contentedly they return to dust again, and are laid low, resigned to lie and decay at the foot of the tree, and afford nourishment to new generations of their kind, as well as to flutter on high! They teach us how to die. One wonders if the time will ever come when men, with their boasted faith in immortality, will lie down as gracefully and as ripe, — with such an Indian-summer serenity will shed their bodies, as they do their hair and nails.

When the leaves fall, the whole earth is a cemetery pleasant to walk in. I love to wander and muse over them in their graves. Here are no lying nor vain epitaphs. What though you own no lot at Mount Auburn? Your lot is surely cast somewhere in this vast cemetery, which has been consecrated from of old. You need attend no auction to secure a place. There is room enough here. The loosestrife shall bloom and the huckleberry-bird sing over your bones. The woodman and hunter shall be your sextons, and the children shall tread upon the borders as much as they will. Let us walk in the cemetery of the leaves; this is your true Greenwood Cemetery.

THE SUGAR MAPLE

But think not that the splendor of the year is over; for as one leaf does not make a summer, neither does one falling leaf make an autumn. The smallest sugar maples in our streets make a great show as early as the fifth of October, more than any other trees there. As I look up the main street, they appear like painted screens standing before the houses; yet many are green. But now, or generally by the seventeenth of October, when almost all red maples and some white maples are bare, the large sugar maples also are in their glory, glowing with yellow and red, and show unexpectedly bright and delicate tints. They are remarkable for the contrast they often afford of deep blushing red on one half and green on the other. They become at length dense masses of rich yellow with a deep scarlet blush, or more than blush, on the exposed surfaces. They are the brightest trees now in the street.

The large ones on our Common are particularly beautiful. A delicate but warmer than golden yellow is now the prevailing color, with scarlet cheeks. Yet, standing on the east side of the Common just before sundown, when the western light is transmitted through them, I see that their yellow even, compared with the pale lemon yellow of an elm close by, amounts to a scarlet, without noticing the bright scarlet portions. Generally, they are great regular oval masses of yellow and scarlet. All the sunny warmth of the season, the Indian summer, seems to be absorbed in their leaves. The lowest and inmost leaves next the bole are, as usual, of the most delicate yellow and green, like the complexion of young men brought up in the house. There is an auction on the Common to-day, but its red flag is hard to be discerned amid this blaze of color.

Little did the fathers of the town anticipate this brilliant success, when they caused to be imported from farther in the country some straight poles with their tops cut off, which they called sugar maples; and, as I remember, after they were set out, a neighboring merchant's clerk, by way of jest, planted beans about them. Those which were then jestingly called bean-poles are to-day far the most beautiful objects noticeable in our streets. They are worth all and more than they have cost, — though one of the selectmen, while setting them out, took the cold which occasioned his death, — if only because they have filled the open eyes of children with their rich color unstintedly so many Octobers. We will not ask them to yield us sugar in the spring, while they afford us so fair a prospect in the autumn. Wealth indoors may be the inheritance of few, but it is equally distributed on the Common. All children alike can revel in this golden harvest.

Surely trees should be set in our streets with a view to their October splendor, though I doubt whether this is ever considered by the "Tree Society." Do you not think it will make some odds to these children that they were brought up under the maples? Hundreds of eyes are steadily drinking in this color, and by these teachers even the truants are caught and educated the moment they step abroad. Indeed, neither the truant nor the studious is at present taught color in the schools.

These are instead of the bright colors in apothecaries' shops and city windows. It is a pity that we have no more *red* maples, and some hickories, in our streets as well. Our paint-box is very imperfectly filled. Instead of, or beside, supplying such paint-boxes as we do, we might supply these natural colors to the young. Where else will they study color under greater advantages? What School of Design can vie with this? Think how much the eyes of painters of all kinds, and of manufacturers of cloth and paper, and paper-stainers, and countless others, are to be educated by these autumnal colors. The stationer's envelopes may be of very various tints, yet not so various as those of the leaves of a single tree. If you want a different shade or tint of a particular color, you have only to look farther within or without the tree or the wood. These leaves are not many dipped in one dye, as at the dye-house, but they are dyed in light of infinitely various degrees of strength and left to set and dry there.

Shall the names of so many of our colors continue to be derived from those of obscure foreign localities, as Naples yellow, Prussian blue, raw Sienna, burnt Umber, Gamboge? (surely the Tyrian purple must have faded by this time), or from comparatively trivial articles of commerce, — chocolate, lemon, coffee, cinnamon, claret? (shall we compare our hickory to a lemon, or a lemon to a hickory?) or from ores and oxides which few ever see? Shall we so often, when describing to our neighbors the color of something we have seen, refer them, not to some natural object in our neighborhood, but perchance to a bit of earth fetched from the other side of the planet, which possibly they may find at the apothecary's, but which probably neither they nor we ever saw? Have we not an *earth* under our feet, — aye, and a sky over our heads? Or is the last *all* ultramarine? What do we know of sapphire, amethyst, emerald, ruby, amber, and the like, — most of us who take these names in vain? Leave these precious words to cabinet-keepers, virtuosos, and maids-of-honor, — to the Nabobs, Begums, and Chobdars of Hindostan, or wherever else. I do not see why, since America and her autumn woods have been discovered, our leaves should not compete with the precious stones in giving names to colors; and, indeed, I believe that in course of time the names of some of our trees and shrubs, as well as flowers, will get into our popular chromatic nomenclature.

But of much more importance than a knowledge of the names and distinctions of color is the joy and exhilaration which these colored leaves excite. Already these brilliant trees throughout the street, without any more variety, are at least equal to an annual festival and holiday, or a week of such. These are cheap and innocent gala-days, celebrated by one and all without the aid of committees or marshals, such a show as may safely be licensed, not attracting gamblers or rum-sellers, not requiring any special police to keep the peace. And poor indeed must be that New England village's October which has not the maple in its streets. This October festival costs no powder, nor ringing of bells, but every tree is a living liberty-pole on which a thousand bright flags are waving.

No wonder that we must have our annual cattle-show, and fall training, and perhaps cornwallis, our September courts, and the like. Nature herself holds her annual fair in October, not only in the streets, but in every hollow and on every hillside. When lately we looked into that red maple swamp all ablaze, where the trees were clothed in their vestures of most dazzling tints, did it not suggest a thousand gypsies beneath, — a race capable of wild delight, — or even the fabled fauns, satyrs, and wood-nymphs come back to earth? Or was it only a congregation of wearied woodchoppers, or of proprietors come to inspect their lots, that we thought of? Or, earlier still, when we paddled on the river through that fine-grained September air, did there not appear to be something new going on under the sparkling surface of the stream, a shaking of props, at least, so that we made haste in order to be up in time? Did not the rows of yellowing willows and button-bushes on each side seem like rows of booths, under which, perhaps, some fluviatile egg-pop equally yellow was effervescing? Did not all these suggest that man's spirits should rise as high as Nature's, — should hang out their flag, and the routine of his life be interrupted by an analogous expression of joy and hilarity?

No annual training or muster of soldiery, no celebration with its scarfs and banners, could import into the town a hundredth part of the annual splendor of our October. We have only to set the trees, or let them stand, and Nature will find the colored drapery, — flags of all her nations, some of whose private signals hardly the botanist can read, — while we walk under the triumphal arches of the elms. Leave it to Nature to appoint the days, whether the same as in neighboring States or not, and let the clergy read her proclamations, if they can understand them. Behold what a brilliant drapery is her woodbine flag! What public-spirited merchant, think you, has contributed this part of the show? There is no handsomer shingling and paint than this vine, at present covering a whole side of some houses. I do not believe that the ivy *never sere* is comparable to it. No wonder it has been extensively introduced into London. Let us have a good many maples and hickories and scarlet oaks, then, I say. Blaze away! Shall that dirty roll of bunting in the gun-house be all the colors a village can display? A village is not complete, unless it have these trees to mark the season in it. They are important, like the town clock. A village that has them not will not be found to work well. It has a screw loose, an essential part is wanting. Let us have willows for spring, elms for summer, maples and walnuts and tupeloes for autumn, evergreens for winter, and oaks for all seasons. What is a gallery in a house to a gallery in the streets, which every market-man rides through, whether he will or not? Of course, there is not a picture-gallery in the country which would be worth so much to us as is the western view at sunset under the elms of our main street. They are the frame to a picture which is daily painted behind them. An avenue of elms as large as our largest and three miles long would seem to lead to some admirable place, though only C—— were at the end of it.

A village needs these innocent stimulants of bright

and cheering prospects to keep off melancholy and superstition. Show me two villages, one embowered in trees and blazing with all the glories of October, the other a merely trivial and treeless waste, or with only a single tree or two for suicides, and I shall be sure that in the latter will be found the most starved and bigoted religionists and the most desperate drinkers. Every wash-tub and milk-can and gravestone will be exposed. The inhabitants will disappear abruptly behind their barns and houses, like desert Arabs amid their rocks, and I shall look to see spears in their hands. They will be ready to accept the most barren and forlorn doctrine, — as that the world is speedily coming to an end, or has already got to it, or that they themselves are turned wrong side outward. They will perchance crack their dry joints at one another and call it a spiritual communication.

But to confine ourselves to the maples. What if we were to take half as much pains in protecting them as we do in setting them out, — not stupidly tie our horses to our dahlia stems?

What meant the fathers by establishing this *perfectly living* institution before the church, — this institution which needs no repairing nor repainting, which is continually enlarged and repaired by its growth? Surely they

> "Wrought in a sad sincerity;
> Themselves from God they could not free;
> They *planted* better than they knew; —
> The conscious *trees* to beauty grew."

Verily these maples are cheap preachers, permanently

settled, which preach their half-century, and century, aye, and century-and-a-half sermons, with constantly increasing unction and influence, ministering to many generations of men; and the least we can do is to supply them with suitable colleagues as they grow infirm.

THE SCARLET OAK

Belonging to a genus which is remarkable for the beautiful form of its leaves, I suspect that some scarlet oak leaves surpass those of all other oaks in the rich and wild beauty of their outlines. I judge from an acquaintance with twelve species, and from drawings which I have seen of many others.

Stand under this tree and see how finely its leaves are cut against the sky, — as it were, only a few sharp points extending from a midrib. They look like double, treble, or quadruple crosses. They are far more ethereal than the less deeply scalloped oak leaves. They have so little leafy *terra firma* that they appear melting away in the light, and scarcely obstruct our view. The leaves of very young plants are, like those of full-grown oaks of other species, more entire, simple, and lumpish in their outlines, but these, raised high on old trees, have solved the leafy problem. Lifted higher and higher, and sublimated more and more, putting off some earthiness and cultivating more intimacy with the light each year, they have at length the least possible amount of earthy matter, and the greatest spread and grasp of skyey influences. There they dance, arm in arm with the light, — tripping it on fantastic points, fit partners in those aerial halls. So intimately mingled are they with it, that, what with their slenderness and their glossy surfaces, you can hardly tell at last what in the dance is leaf and what is light. And when no zephyr stirs, they are at most but a rich tracery to the forest windows.

I am again struck with their beauty, when, a month later, they thickly strew the ground in the woods, piled one upon another under my feet. They are then brown above, but purple beneath. With their narrow lobes and their bold, deep scallops reaching almost to the middle, they suggest that the material must be cheap, or else there has been a lavish expense in their creation, as if so much had been cut out. Or else they seem to us the remnants of the stuff out of which leaves have been cut with a die. Indeed, when they lie thus one upon another, they remind me of a pile of scrap-tin.

Or bring one home, and study it closely at your leisure, by the fireside. It is a type, not from any Oxford font, not in the Basque nor the arrow-headed character, not found on the Rosetta Stone, but destined to be copied in sculpture one day, if they ever get to whittling stone here. What a wild and pleasing outline, a combination of graceful curves and angles! The eye rests with equal delight on what is not leaf and on what is leaf, — on the broad, free, open sinuses, and on the long, sharp, bristle-pointed lobes. A simple oval outline would include it all, if you connected the points of the leaf; but how much richer is it than that, with its half-dozen deep scallops, in which the eye and thought of the beholder are embayed! If I were a drawing-master, I would set my pupils to copying these leaves, that they might learn to draw firmly and gracefully.

Regarded as water, it is like a pond with half a dozen broad rounded promontories extending nearly to its middle, half from each side, while its watery bays extend far inland, like sharp friths, at each of whose heads several fine streams empty in, — almost a leafy archipelago.

But it oftener suggests land, and, as Dionysius and Pliny compared the form of the Morea to that of the leaf of the Oriental plane tree, so this leaf reminds me of some fair wild island in the ocean, whose extensive coast, alternate rounded bays with smooth strands, and sharp-pointed rocky capes, mark it as fitted for the habitation of man, and destined to become a centre of civilization at last. To the sailor's eye, it is a much indented shore. Is it not, in fact, a shore to the aerial ocean, on which the windy surf beats? At sight of this leaf we are all mariners, — if not vikings, buccaneers, and filibusters. Both our love of repose and our spirit of adventure are addressed. In our most casual glance, perchance, we think that if we succeed in doubling those sharp capes we shall find deep, smooth, and secure havens in the ample bays. How different from the white oak leaf, with its rounded headlands, on which no lighthouse need be placed! That is an England, with its long civil history, that may be read. This is some still unsettled New-found Island or Celebes. Shall we go and be rajahs there?

By the twenty-sixth of October the large scarlet oaks are in their prime, when other oaks are usually withered. They have been kindling their fires for a week past, and now generally burst into a blaze. This alone of *our* indigenous deciduous trees (excepting the dogwood, of which I do not know half a dozen, and they are but large bushes) is now in its glory. The two aspens and the sugar maple come nearest to it in date, but they have lost the greater part of their leaves. Of evergreens, only the pitch pine is still commonly bright.

But it requires a particular alertness, if not devotion to these phenomena, to appreciate the wide-spread, but late and unexpected glory of the scarlet oaks. I do not speak here of the small trees and shrubs, which are commonly observed, and which are now withered, but of the large trees. Most go in and shut their doors, thinking that bleak and colorless November has already come, when some of the most brilliant and memorable colors are not yet lit.

This very perfect and vigorous one, about forty feet high, standing in an open pasture, which was quite glossy green on the twelfth, is now, the twenty-sixth, completely changed to bright dark-scarlet, — every leaf, between you and the sun, as if it had been dipped into a scarlet dye. The whole tree is much like a heart in form, as well as color. Was not this worth waiting for? Little did you think, ten days ago, that that cold green tree would assume such color as this. Its leaves are still firmly attached, while those of other trees are falling around it. It seems to say: "I am the last to blush, but I blush deeper than any of ye. I bring up the rear in my red coat. We scarlet ones, alone of oaks, have not given up the fight."

The sap is now, and even far into November, frequently flowing fast in these trees, as in maples in the spring; and apparently their bright tints, now that most other oaks are withered, are connected with this phenomenon. They are full of life. It has a pleasantly astringent, acorn-like taste, this strong oak wine, as I find on tapping them with my knife.

Looking across this woodland valley, a quarter of a mile wide, how rich those scarlet oaks embosomed in pines, their bright red branches intimately intermingled with them! They have their full effect there. The pine boughs are the green calyx to their red petals. Or, as we go along a road in the woods, the sun striking endwise through it, and lighting up the red tents of the oaks, which on each side are mingled with the liquid green of the pines, makes a very gorgeous scene. Indeed, without the evergreens for contrast, the autumnal tints would lose much of their effect.

The scarlet oak asks a clear sky and the brightness of late October days. These bring out its colors. If the sun goes into a cloud they become comparatively indistinct. As I sit on a cliff in the southwest part of our town, the sun is now getting low, and the woods in Lincoln, south and east of me, are lit up by its more level rays; and in the scarlet oaks, scattered so equally over the forest, there is brought out a more brilliant redness than I had believed was in them. Every tree of this species which is visible in those directions, even to the horizon, now stands out distinctly red. Some great ones lift their red backs high above the woods, in the next town, like huge roses with a myriad of fine petals; and some more slender ones, in a small grove of white pines on Pine Hill in the east, on the very verge of the horizon, alternating with the pines on the edge of the grove, and shouldering them with their red coats, look like soldiers in red amid hunters in green. This time it is Lincoln green, too. Till the sun got low, I did not believe that there were so many redcoats in the forest army. Theirs is an intense, burning red, which would lose some of its strength, methinks, with every step you might take toward them; for the shade that lurks amid their foliage does not report itself at this distance, and they are unanimously red. The focus of their reflected color is in the atmosphere far on this side. Every such tree becomes a nucleus of red, as it were, where, with the declining sun, that color grows and glows. It is partly borrowed fire, gathering strength from the sun on its way to your eye. It has only some comparatively dull red leaves for a rallying-point, or kindling-stuff, to start it, and it becomes an intense scarlet or red mist, or fire, which finds fuel for itself in the very atmosphere. So vivacious is redness. The very rails reflect a rosy light at this hour and season. You see a redder tree than exists.

If you wish to count the scarlet oaks, do it now. In a clear day stand thus on a hilltop in the woods, when the sun is an hour high, and every one within range of your vision, excepting in the west, will be revealed. You might live to the age of Methuselah and never find a tithe of them, otherwise. Yet sometimes even in a dark day I have thought them as bright as I ever saw them. Looking westward, their colors are lost in a blaze of light; but in other directions the whole forest is a flower-garden, in which these late roses burn, alternating with green, while the so-called "gardeners," walking here and there, perchance, beneath, with spade and waterpot, see only a few little asters amid withered leaves.

These are *my* China-asters, *my* late garden-flowers. It costs me nothing for a gardener. The falling leaves, all over the forest, are protecting the roots of my plants. Only look at what is to be seen, and you will have garden enough, without deepening the soil in your yard. We have only to elevate our view a little, to see the whole forest as a garden. The blossoming of the scarlet oak, — the forest-flower, surpassing all in splendor (at least since the maple)! I do not know but they interest me more than the maples, they are so widely and equally dispersed throughout the forest; they are so hardy, a nobler tree on the whole; our chief November flower, abiding the approach of winter with us, imparting warmth to early November prospects. It is remarkable that the latest bright color that is general should be this deep, dark scarlet and red, the intensest of colors. The ripest fruit of the year; like the cheek of a hard, glossy red apple, from the cold Isle of Orleans, which will not be mellow for eating till next spring! When I rise to a hilltop, a thousand of these great oak roses, distributed on every side, as far as the horizon! I admire them four or five miles off! This my unfailing prospect for a fortnight past! This late forest-flower surpasses all that spring or summer could do. Their colors were but rare and dainty specks comparatively (created for the near-sighted, who walk amid the humblest herbs and underwoods), and made no impression on a distant eye. Now it is an extended forest or a mountain-side, through or along which we journey from day to day, that bursts into bloom. Comparatively, our gardening is on a petty scale, — the gardener still nursing a few asters amid dead weeds, ignorant of the gigantic asters and roses which, as it were, overshadow him, and ask for none of his care. It is like a little red paint ground on a saucer, and held up against the sunset sky. Why not take more elevated and broader views, walk in the great garden; not skulk in a little "debauched" nook of it? consider the beauty of the forest, and not merely of a few impounded herbs?

Let your walks now be a little more adventurous; ascend the hills. If, about the last of October, you ascend any hill in the outskirts of our town, and probably of yours, and look over the forest, you may see — well, what I have endeavored to describe. All this you surely *will* see, and much more, if you are prepared to see it, — if you *look* for it. Otherwise, regular and universal as this phenomenon is, whether you stand on the hilltop or in the hollow, you will think for threescore years and ten that all the wood is, at this season, sere and brown. Objects are concealed from our view, not so much because they are out of the course of our visual ray as because we do not bring our minds and eyes to bear on them; for there is no power to see in the eye itself, any more than in any other jelly. We do not realize how far and widely, or how near and narrowly, we are to look. The greater part of the phenomena of Nature are for this reason concealed from us all our lives. The gardener sees only the gardener's garden. Here, too, as in political economy, the supply answers to the demand. Nature does not cast pearls before swine. There is just as much beauty visible to us in the landscape as we are prepared to appreciate, — not a grain more. The actual objects which one man will see from a particular hilltop are just as different from those which another will see as the beholders are different. The scarlet oak must, in a sense, be in your eye when you go forth. We cannot see anything until we are possessed with the idea of it, take it into our heads, — and then we can hardly see anything else. In my botanical rambles I find that, first, the idea, or image, of a plant occupies my thoughts, though it may seem very foreign to this locality, — no nearer than Hudson's Bay, — and for some weeks or months I go thinking of it, and expecting it, unconsciously, and at length I surely see it. This is the history of my finding a score or more of rare plants which I could name. A man sees only what concerns him. A botanist absorbed in the study of grasses does not distinguish the grandest pasture oaks. He, as it were, tramples down oaks unwittingly in his walk, or at most sees only their shadows. I have found that it required a different intention of the eye, in the same locality, to see different plants, even when they were closely allied, as *Juncaceae* and *Gramineae:* when I was looking for the former, I did not see the latter in the midst of them. How much more, then, it requires different intentions of the eye and of the mind to attend to different departments of knowledge! How differently the poet and the naturalist look at objects!

Take a New England selectman, and set him on the highest of our hills, and tell him to look, — sharpening his sight to the utmost, and putting on the glasses that suit him best (aye, using a spy-glass, if he likes), — and make a full report. What, probably, will he *spy?* — what will he *select* to look at? Of course, he will see a Brocken spectre of himself. He will see several meeting-houses, at least, and, perhaps, that somebody ought to be assessed higher than he is, since he has so handsome a wood-lot. Now take Julius Cæsar, or Emanuel Swedenborg, or a Fiji-Islander, and set him up there. Or suppose all together, and let them compare notes afterward. Will it appear that they have enjoyed the same prospect? What they will see will be as different as Rome was from heaven or hell, or the last from the Fiji Islands. For aught we know, as strange a man as any of these is always at our elbow.

Why, it takes a sharpshooter to bring down even such trivial game as snipes and woodcocks; he must take very particular aim, and know what he is aiming at. He would stand a very small chance, if he fired at random into the sky, being told that snipes were flying there. And so is it with him that shoots at beauty; though he wait till the sky falls, he will not bag any, if he does not already know its seasons and haunts, and the color of its wing, — if he has not dreamed of it, so that he can *anticipate* it; then, indeed, he flushes it at every step, shoots double and on the wing, with both barrels, even in corn-fields. The sportsman trains himself, dresses, and watches unweariedly, and loads and primes for his particular game. He prays for it, and offers sacrifices, and so he gets it. After due and long preparation, schooling his eye and hand, dreaming awake and asleep, with gun and paddle and boat, he goes out after meadow-hens, which most of his townsmen never saw nor dreamed of, and paddles for miles against a head wind, and wades in water up to his knees, being out all day without his dinner, and *therefore* he gets them. He had them half-way into his bag when he started, and has only to shove them down. The true sportsman can shoot you almost any of his game from his windows: what else has he windows or eyes for? It comes and perches at last on the barrel of his gun; but the rest of the world never see it *with the feathers on.* The geese fly exactly under his zenith, and honk when they get there, and he will keep himself supplied by firing up his chimney; twenty musquash have the refusal of each one of his traps before it is empty. If he lives, and his game spirit increases, heaven and earth shall fail him sooner than game; and when he dies, he will go to more extensive and, perchance, happier hunting-grounds. The fisherman, too, dreams of fish, sees a bobbing cork in his dreams, till he can almost catch them in his sink-spout. I knew a girl who, being sent to pick huckleberries, picked wild gooseberries by the quart, where no one else knew that there were any, because she was accustomed to pick them up-country where she came from. The astronomer knows where to go star-gathering, and sees one clearly in his mind before any have seen it with a glass. The hen scratches and finds her food right under where she stands; but such is not the way with the hawk.

These bright leaves which I have mentioned are not the exception, but the rule; for I believe that all leaves, even grasses and mosses, acquire brighter colors just before their fall. When you come to observe faithfully the changes of each humblest plant, you find that each has, sooner or later, its peculiar autumnal tint; and if you undertake to make a complete list of the bright tints, it will be nearly as long as a catalogue of the plants in your vicinity.

WILD APPLES

THE HISTORY OF THE APPLE TREE

It is remarkable how closely the history of the apple tree is connected with that of man. The geologist tells us that the order of the *Rosaceae*, which includes the apple, also the true grasses, and the *Labiatae*, or mints, were introduced only a short time previous to the appearance of man on the globe.

It appears that apples made a part of the food of that unknown primitive people whose traces have lately been found at the bottom of the Swiss lakes, supposed to be older than the foundation of Rome, so old that they had no metallic implements. An entire black and shriveled crab-apple has been recovered from their stores.

Tacitus says of the ancient Germans that they satisfied their hunger with wild apples (*agrestia poma*), among other things.

Niebuhr observes that "the words for a house, a field, a plow, plowing, wine, oil, milk, sheep, apples, and others relating to agriculture and the gentler way of life, agree in Latin and Greek, while the Latin words for all objects pertaining to war or the chase are utterly alien from the Greek." Thus the apple tree may be considered a symbol of peace no less than the olive.

The apple was early so important, and generally distributed, that its name traced to its root in many languages signifies fruit in general. Μῆλον, in Greek, means an apple, also the fruit of other trees, also a sheep and any cattle, and finally riches in general.

The apple tree has been celebrated by the Hebrews, Greeks, Romans, and Scandinavians. Some have thought that the first human pair were tempted by its fruit. Goddesses are fabled to have contended for it, dragons were set to watch it, and heroes were employed to pluck it.

The tree is mentioned in at least three places in the Old Testament, and its fruit in two or three more. Solomon sings, "As the apple-tree among the trees of the wood, so is my beloved among the sons." And again, "Stay me with flagons, comfort me with apples." The noblest part of man's noblest feature is named from this fruit, "the apple of the eye."

The apple tree is also mentioned by Homer and Herodotus. Ulysses saw in the glorious garden of Alcinoüs "pears and pomegranates, and apple trees bearing beautiful fruit" (καὶ μηλέαι ἀγλαόκαρποι). And according to Homer, apples were among the fruits which Tantalus could not pluck, the wind ever blowing their boughs away from him. Theophrastus knew and described the apple tree as a botanist.

According to the Prose Edda, "Iduna keeps in a box the apples which the gods, when they feel old age approaching, have only to taste of to become young again. It is in this manner that they will be kept in renovated youth until Ragnarök" (or the destruction of the gods).

I learn from Loudon that "the ancient Welsh bards were rewarded for excelling in song by the token of the apple-spray;" and "in the Highlands of Scotland the apple-tree is the badge of the clan Lamont."

The apple tree (*Pyrus malus*) belongs chiefly to the northern temperate zone. Loudon says that "it grows spontaneously in every part of Europe except the frigid zone, and throughout Western Asia, China, and Japan." We have also two or three varieties of the apple indigenous in North America. The cultivated apple tree was first introduced into this country by the earliest settlers, and is thought to do as well or better here than anywhere else. Probably some of the varieties which are now cultivated were first introduced into Britain by the Romans.

Pliny, adopting the distinction of Theophrastus, says, "Of trees there are some which are altogether wild (*sylvestres*), some more civilized (*urbaniores*)." Theophrastus includes the apple among the last; and, indeed, it is in this sense the most civilized of all trees. It is as harmless as a dove, as beautiful as a rose, and as valuable as flocks and herds. It has been longer cultivated than any other, and so is more humanized; and who knows but, like the dog, it will at length be no longer traceable to its wild original? It migrates with man, like the dog and horse and cow: first, perchance, from Greece to Italy, thence to England, thence to America; and our Western emigrant is still marching steadily toward the setting sun with the seeds of the apple in his pocket, or perhaps a few young trees strapped to his load. At least a million apple trees are thus set farther westward this year than any cultivated ones grew last year. Consider how the Blossom Week, like the Sabbath, is thus annually spreading over the prairies; for when man migrates, he carries with him not only his birds, quadrupeds, insects, vegetables, and his very sward, but his orchard also.

The leaves and tender twigs are an agreeable food to many domestic animals, as the cow, horse, sheep, and goat; and the fruit is sought after by the first, as well as by the hog. Thus there appears to have existed a natural alliance between these animals and this tree from the first. "The fruit of the crab in the forests of France" is said to be "a great resource for the wild boar."

Not only the Indian, but many indigenous insects, birds, and quadrupeds, welcomed the apple tree to these shores. The tent caterpillar saddled her eggs on the very first twig that was formed, and it has since shared her affections with the wild cherry; and the canker-worm also in a measure abandoned the elm to feed on it. As it grew apace, the bluebird, robin, cherry-bird, kingbird, and many more came with haste and built their nests and warbled in its boughs, and so became orchard-birds, and multiplied more than ever. It was an era in the history of their race. The downy woodpecker found such a savory morsel under its bark that he perforated it in a ring quite round the tree, before he left it, — a thing which he had never done before, to my knowledge. It did not take the partridge long to find out how sweet its buds were, and every winter eve she flew, and still flies, from the wood, to pluck them, much to the farmer's sorrow. The rabbit, too, was not slow to learn the taste of its twigs and bark;

and when the fruit was ripe, the squirrel half rolled, half carried it to his hole; and even the musquash crept up the bank from the brook at evening, and greedily devoured it, until he had worn a path in the grass there; and when it was frozen and thawed, the crow and the jay were glad to taste it occasionally. The owl crept into the first apple tree that became hollow, and fairly hooted with delight, finding it just the place for him; so, settling down into it, he has remained there ever since.

My theme being the Wild Apple, I will merely glance at some of the seasons in the annual growth of the cultivated apple, and pass on to my special province.

The flowers of the apple are perhaps the most beautiful of any tree's, so copious and so delicious to both sight and scent. The walker is frequently tempted to turn and linger near some more than usually handsome one, whose blossoms are two-thirds expanded. How superior it is in these respects to the pear, whose blossoms are neither colored nor fragrant!

By the middle of July, green apples are so large as to remind us of coddling, and of the autumn. The sward is commonly strewed with little ones which fall stillborn, as it were, — Nature thus thinning them for us. The Roman writer Palladius said, "If apples are inclined to fall before their time, a stone placed in a split root will retain them." Some such notion, still surviving, may account for some of the stones which we see placed, to be overgrown, in the forks of trees. They have a saying in Suffolk, England, —

> "At Michaelmas time, or a little before,
> Half an apple goes to the core."

Early apples begin to be ripe about the first of August; but I think that none of them are so good to eat as some to smell. One is worth more to scent your handkerchief with than any perfume which they sell in the shops. The fragrance of some fruits is not to be forgotten, along with that of flowers. Some gnarly apple which I pick up in the road reminds me by its fragrance of all the wealth of Pomona, — carrying me forward to those days when they will be collected in golden and ruddy heaps in the orchards and about the cider-mills.

A week or two later, as you are going by orchards or gardens, especially in the evenings, you pass through a little region possessed by the fragrance of ripe apples, and thus enjoy them without price, and without robbing anybody.

There is thus about all natural products a certain volatile and ethereal quality which represents their highest value, and which cannot be vulgarized, or bought and sold. No mortal has ever enjoyed the perfect flavor of any fruit, and only the godlike among men begin to taste its ambrosial qualities. For nectar and ambrosia are only those fine flavors of every earthly fruit which our coarse palates fail to perceive, — just as we occupy the heaven of the gods without knowing it. When I see a particularly mean man carrying a load of fair and fragrant early apples to market, I seem to see a contest going on between him and his horse, on the one side, and the apples on the other, and, to my mind, the apples always gain it. Pliny says that apples are the heaviest of all things, and that the oxen begin to sweat at the mere sight of a load of them. Our driver begins to lose his load the moment he tries to transport them to where they do not belong, that is, to any but the most beautiful. Though he gets out from time to time, and feels of them, and thinks they are all there, I see the stream of their evanescent and celestial qualities going to heaven from his cart, while the pulp and skin and core only are going to market. They are not apples, but pomace. Are not these still Iduna's apples, the taste of which keeps the gods forever young? and think you that they will let Loki or Thjassi carry them off to Jötunheim, while they grow wrinkled and gray? No, for Ragnarök, or the destruction of the gods, is not yet.

There is another thinning of the fruit, commonly near the end of August or in September, when the ground is strewn with windfalls; and this happens especially when high winds occur after rain. In some orchards you may see fully three quarters of the whole crop on the ground, lying in a circular form beneath the trees, yet hard and green, or, if it is a hillside, rolled far down the hill. However, it is an ill wind that blows nobody any good. All the country over, people are busy picking up the windfalls, and this will make them cheap for early apple pies.

In October, the leaves falling, the apples are more distinct on the trees. I saw one year in a neighboring town some trees fuller of fruit than I remember to have ever seen before, small yellow apples hanging over the road. The branches were gracefully drooping with their weight, like a barberry bush, so that the whole tree acquired a new character. Even the topmost branches, instead of standing erect, spread and drooped in all directions; and there were so many poles supporting the lower ones that they looked like pictures of banyan trees. As an old English manuscript says, "The mo appelen the tree bereth the more sche boweth to the folk."

Surely the apple is the noblest of fruits. Let the most beautiful or the swiftest have it. That should be the "going" price of apples.

Between the 5th and 20th of October I see the barrels lie under the trees. And perhaps I talk with one who is selecting some choice barrels to fulfill an order. He turns a specked one over many times before he leaves it out. If I were to tell what is passing in my mind, I should say that every one was specked which he had handled; for he rubs off all the bloom, and those fugacious ethereal qualities leave it. Cool evenings prompt the farmers to make haste, and at length I see only the ladders here and there left leaning against the trees.

It would be well, if we accepted these gifts with more joy and gratitude, and did not think it enough simply to put a fresh load of compost about the tree. Some old English customs are suggestive at least. I find them described chiefly in Brand's "Popular Antiquities." It appears that "on Christmas Eve the farmers and their men in Devonshire take a large bowl of cider, with a toast in it, and carrying it in state to the orchard, they salute the apple-trees with much ceremony, in order to make them bear well the next season." This salutation consists in "throwing some of the cider about the roots of the tree, placing bits of the toast on the branches," and then, "encircling one of the best bearing trees in the orchard, they drink the following toast three several times: —

> 'Here's to thee, old apple tree,
> Whence thou mayst bud, and whence thou mayst blow,
> And whence thou mayst bear apples enow!
> Hats-full! caps-full!
> Bushel, bushel, sacks-full!
> And my pockets full, too! Hurra!'"

Also what was called "apple-howling" used to be practiced in various counties of England on New Year's Eve. A troop of boys visited the different orchards, and, encircling the apple trees, repeated the following words: —

> "Stand fast, root! bear well, top!
> Pray God send us a good howling crop:
> Every twig, apples big;
> Every bough, apples enow!"

"They then shout in chorus, one of the boys accompanying them on a cow's horn. During this ceremony they rap the trees with their sticks." This is called "wassailing" the trees, and is thought by some to be "a relic of the heathen sacrifice to Pomona."

Herrick sings, —

> "Wassaile the trees that they may beare
> You many a plum and many a peare;
> For more or less fruits they will bring
> As you so give them wassailing."

Our poets have as yet a better right to sing of cider than of wine; but it behooves them to sing better than English Phillips did, else they will do no credit to their Muse.

THE WILD APPLE

So much for the more civilized apple trees (*urbaniores*, as Pliny calls them). I love better to go through the old orchards of ungrafted appletrees, at what ever season of the year, — so irregularly planted: sometimes two trees standing close together; and the rows so devious that you would think that they not only had grown while the owner was sleeping, but had been set out by him in a somnambulic state. The rows of grafted fruit will never tempt me to wander amid them like these. But I now, alas, speak rather from memory than from any recent experience, such ravages have been made!

Some soils, like a rocky tract called the Easterbrooks Country in my neighborhood, are so suited to the apple, that it will grow faster in them without any care, or if only the ground is broken up once a year, than it will in many places with any amount of care. The owners of this tract allow that the soil is excellent for fruit, but they say that it is so rocky that they have not patience to plow it, and that, together with the distance, is the reason why it is not cultivated. There are, or were recently, extensive orchards there standing without order. Nay, they spring up wild and bear well there in the midst of pines, birches, maples, and oaks. I am often surprised to see rising amid these trees the rounded tops of apple trees glowing with red or yellow fruit, in harmony with the autumnal tints of the forest.

Going up the side of a cliff about the first of November, I saw a vigorous young apple tree, which, planted by birds or cows, had shot up amid the rocks and open woods there, and had now much fruit on it, uninjured by the frosts, when all cultivated apples were gathered. It was a rank, wild growth, with many green leaves on it still, and made an impression of thorniness. The fruit was hard and green, but looked as if it would be palatable in the winter. Some was dangling on the twigs, but more half buried in the wet leaves under the tree, or rolled far down the hill amid the rocks. The owner knows nothing of it. The day was not observed when it first blossomed, nor when it first bore fruit, unless by the chickadee. There was no dancing on the green beneath it in its honor, and now there is no hand to pluck its fruit, — which is only gnawed by squirrels, as I perceive. It has done double duty, — not only borne this crop, but each twig has grown a foot into the air. And this is *such* fruit! bigger than many berries, we must admit, and carried home will be sound and palatable next spring. What care I for Iduna's apples so long as I can get these?

When I go by this shrub thus late and hardy, and see its dangling fruit, I respect the tree, and I am grateful for Nature's bounty, even though I cannot eat it. Here on this rugged and woody hillside has grown an apple tree, not planted by man, no relic of a former orchard, but a natural growth, like the pines and oaks. Most fruits which we prize and use depend entirely on our care. Corn and grain, potatoes, peaches, melons, etc., depend altogether on our planting; but the apple emulates man's independence and enterprise. It is not simply carried, as I have said, but, like him, to some extent, it has migrated to this New World, and is even, here and there, making its way amid the aboriginal trees; just as the ox and dog and horse sometimes run wild and maintain themselves.

Even the sourest and crabbedest apple, growing in the most unfavorable position, suggests such thoughts as these, it is so noble a fruit.

THE CRAB

Nevertheless, *our* wild apple is wild only like myself, perchance, who belong not to the aboriginal race here, but have strayed into the woods from the cultivated stock. Wilder still, as I have said, there grows elsewhere in this country a native and aboriginal crab-apple, *Malus coronaria*, "whose nature has not yet been modified by cultivation." It is found from western New York to Minnesota, and southward. Michaux says that its ordinary height "is fifteen or eighteen feet, but it is sometimes found twenty-five or thirty feet high," and that the large ones "exactly resemble the common apple tree." "The flowers are white mingled with rose color, and are collected in corymbs." They are remarkable for their delicious odor. The fruit, according to him, is about an inch and a half in diameter, and is intensely acid. Yet they make fine sweetmeats and also cider of them. He concludes that "if, on being cultivated, it does not yield new and palatable varieties, it will at least be celebrated for the beauty of its flowers, and for the sweetness of its perfume."

I never saw the crab-apple till May, 1861. I had heard of it through Michaux, but more modern botanists, so far as I know, have not treated it as of any peculiar importance. Thus it was a half-fabulous tree to me. I contemplated a pilgrimage to the "Glades," a portion of Pennsylvania where it was said to grow to perfection. I thought of sending to a nursery for it, but doubted if they had it, or would distinguish it from European varieties. At last I had occasion to go to Minnesota, and on entering Michigan I began to notice from the cars a tree with handsome rose-colored flowers. At first I thought it some variety of thorn; but it was not long before the truth flashed on me, that this was my long-sought crab-apple. It was the prevailing flowering shrub or tree to be seen from the cars at that season of the year, — about the middle of May. But the cars never stopped before one, and so I was launched on the bosom of the Mississippi without having touched one, experiencing the fate of Tantalus. On arriving at St. Anthony's Falls, I was sorry to be told that I was too far north for the crab-apple. Nevertheless I succeeded in finding it about eight miles west of the Falls; touched it and smelled it, and secured a lingering corymb of flowers for my herbarium. This must have been near its northern limit.

HOW THE WILD APPLE GROWS

But though these are indigenous, like the Indians, I doubt whether they are any hardier than those backwoodsmen among the apple trees, which, though descended from cultivated stocks, plant themselves in distant fields and forests, where the soil is favorable to them. I know of no trees which have more difficulties to contend with, and which more sturdily resist their foes. These are the ones whose story we have to tell. It oftentimes reads thus: —

Near the beginning of May, we notice little thickets of apple trees just springing up in the pastures where cattle have been, — as the rocky ones of our Easterbrooks Country, or the top of Nobscot Hill, in Sudbury. One or two of these, perhaps, survive the drought and other accidents, — their very birthplace defending them against the encroaching grass and some other dangers, at first.

In two years' time 't had thus
  Reached the level of the rocks,
Admired the stretching world,
  Nor feared the wandering flocks.

But at this tender age
  Its sufferings began:
There came a browsing ox
  And cut it down a span.

This time, perhaps, the ox does not notice it amid the grass; but the next year, when it has grown more stout, he recognizes it for a fellow-emigrant from the old country, the flavor of whose leaves and twigs he well knows; and though at first he pauses to welcome it, and express his surprise, and gets for answer, "The same cause that brought you here brought me," he nevertheless browses it again, reflecting, it may be, that he has some title to it.

Thus cut down annually, it does not despair; but, putting forth two short twigs for every one cut off, it spreads out low along the ground in the hollows or between the rocks, growing more stout and scrubby, until it forms, not a tree as yet, but a little pyramidal, stiff, twiggy mass, almost as solid and impenetrable as a rock. Some of the densest and most impenetrable clumps of bushes that I have ever seen, as well on account of the closeness and stubbornness of their branches as of their thorns, have been these wild apple scrubs. They are more like the scrubby fir and black spruce on which you stand, and sometimes walk, on the tops of mountains, where cold is the demon they contend with, than anything else. No wonder they are prompted to grow thorns at last, to defend themselves against such foes. In their thorniness, however, there is no malice, only some malic acid.

The rocky pastures of the tract I have referred to — for they maintain their ground best in a rocky field — are thickly sprinkled with these little tufts, reminding you often of some rigid gray mosses or lichens, and you see thousands of little trees just springing up between them, with the seed still attached to them.

Being regularly clipped all around each year by the cows, as a hedge with shears, they are often of a perfect conical or pyramidal form, from one to four feet high, and more or less sharp, as if trimmed by the gardener's art. In the pastures on Nobscot Hill and its spurs, they make fine dark shadows when the sun is low. They are also an excellent covert from hawks for many small birds that roost and build in them. Whole flocks perch in them at night, and I have seen three robins' nests in one which was six feet in diameter.

No doubt many of these are already old trees, if you

reckon from the day they were planted, but infants still when you consider their development and the long life before them. I counted the annual rings of some which were just one foot high, and as wide as high, and found that they were about twelve years old, but quite sound and thrifty! They were so low that they were unnoticed by the walker, while many of their contemporaries from the nurseries were already bearing considerable crops. But what you gain in time is perhaps in this case, too, lost in power, — that is, in the vigor of the tree. This is their pyramidal state.

The cows continue to browse them thus for twenty years or more, keeping them down and compelling them to spread, until at last they are so broad that they become their own fence, when some interior shoot, which their foes cannot reach, darts upward with joy: for it has not forgotten its high calling, and bears its own peculiar fruit in triumph.

Such are the tactics by which it finally defeats its bovine foes. Now, if you have watched the progress of a particular shrub, you will see that it is no longer a simple pyramid or cone, but that out of its apex there rises a sprig or two, growing more lustily perchance than an orchard-tree, since the plant now devotes the whole of its repressed energy to these upright parts. In a short time these become a small tree, an inverted pyramid resting on the apex of the other, so that the whole has now the form of a vast hour-glass. The spreading bottom, having served its purpose, finally disappears, and the generous tree permits the now harmless cows to come in and stand in its shade, and rub against and redden its trunk, which has grown in spite of them, and even to taste a part of its fruit, and so disperse the seed.

Thus the cows create their own shade and food; and the tree, its hour-glass being inverted, lives a second life, as it were.

It is an important question with some nowadays, whether you should trim young apple trees as high as your nose or as high as your eyes. The ox trims them up as high as he can reach, and that is about the right height, I think.

In spite of wandering kine, and other adverse circumstances, that despised shrub, valued only by small birds as a covert and shelter from hawks, has its blossom week at last, and in course of time its harvest, sincere, though small.

By the end of some October, when its leaves have fallen, I frequently see such a central sprig, whose progress I have watched, when I thought it had forgotten its destiny, as I had, bearing its first crop of small green or yellow or rosy fruit, which the cows cannot get at over the bushy and thorny hedge which surrounds it, and I make haste to taste the new and undescribed variety. We have all heard of the numerous varieties of fruit invented by Van Mons and Knight. This is the system of Van Cow, and she has invented far more and more memorable varieties than both of them.

Through what hardships it may attain to bear a sweet fruit! Though somewhat small, it may prove equal, if not superior, in flavor to that which has grown in a garden, — will perchance be all the sweeter and more palatable for the very difficulties it has had to contend with. Who knows but this chance wild fruit, planted by a cow or a bird on some remote and rocky hillside, where it is as yet unobserved by man, may be the choicest of all its kind, and foreign potentates shall hear of it, and royal societies seek to propagate it, though the virtues of the perhaps truly crabbed owner of the soil may never be heard of, — at least, beyond the limits of his village? It was thus the Porter and the Baldwin grew.

Every wild apple shrub excites our expectation thus, somewhat as every wild child. It is, perhaps, a prince in disguise. What a lesson to man! So are human beings, referred to the highest standard, the celestial fruit which they suggest and aspire to bear, browsed on by fate; and only the most persistent and strongest genius defends itself and prevails, sends a tender scion upward at last, and drops its perfect fruit on the ungrateful earth. Poets and philosophers and statesmen thus spring up in the country pastures, and outlast the hosts of unoriginal men.

Such is always the pursuit of knowledge. The celestial fruits, the golden apples of the Hesperides, are ever guarded by a hundred-headed dragon which never sleeps, so that it is an Herculean labor to pluck them.

This is one, and the most remarkable way in which the wild apple is propagated; but commonly it springs up at wide intervals in woods and swamp, and by the sides of roads, as the soil may suit it, and grows with comparative rapidity. Those which grow in dense woods are very tall and slender. I frequently pluck from these trees a perfectly mild and tamed fruit. As Palladius says, "*Et injussu consternitur ubere mali:*" And the ground is strewn with the fruit of an unbidden apple tree.

It is an old notion that, if these wild trees do not bear a valuable fruit of their own, they are the best stocks by which to transmit to posterity the most highly prized qualities of others. However, I am not in search of stocks, but the wild fruit itself, whose fierce gust has suffered no "inteneration." It is not my

"highest plot
To plant the Bergamot."

THE FRUIT, AND ITS FLAVOR

The time for wild apples is the last of October and the first of November. They then get to be palatable, for they ripen late, and they are still perhaps as beautiful as ever. I make a great account of these fruits, which the farmers do not think it worth the while to gather, — wild flavors of the Muse, vivacious and inspiriting. The farmer thinks that he has better in his barrels, but he is mistaken, unless he has a walker's appetite and imagination, neither of which can he have.

Such as grow quite wild, and are left out till the first of November, I presume that the owner does not mean to gather. They belong to children as wild as themselves, — to certain active boys that I know, — to the wild-eyed woman of the fields, to whom nothing comes amiss, who gleans after all the world, and, moreover, to us walkers. We have met with them, and they are ours. These rights, long enough insisted upon, have come to be an institution in some old countries, where they have learned how to live. I hear that "the custom of grippling, which may be called apple-gleaning, is, or was formerly, practiced in Herefordshire. It consists in leaving a few apples, which are called the gripples, on every tree, after the general gathering, for the boys, who go with climbing-poles and bags to collect them."

As for those I speak of, I pluck them as a wild fruit, native to this quarter of the earth, — fruit of old trees that have been dying ever since I was a boy and are not yet dead, frequented only by the woodpecker and the squirrel, deserted now by the owner, who has not faith enough to look under their boughs. From the appearance of the tree-top, at a little distance, you would expect nothing but lichens to drop from it, but your faith is rewarded by finding the ground strewn with spirited fruit, — some of it, perhaps, collected at squirrel-holes, with the marks of their teeth by which they carried them, — some containing a cricket or two silently feeding within, and some, especially in damp days, a shell-less snail. The very sticks and stones lodged in the tree-top might have convinced you of the savoriness of the fruit which has been so eagerly sought after in past years.

I have seen no account of these among the "Fruits and Fruit-Trees of America," though they are more memorable to my taste than the grafted kinds; more racy and wild American flavors do they possess when October and November, when December and January, and perhaps February and March even, have assuaged them somewhat. An old farmer in my neighborhood, who always selects the right word, says that "they have a kind of bow-arrow tang."

Apples for grafting appear to have been selected commonly, not so much for their spirited flavor, as for their mildness, their size, and bearing qualities, — not so much for their beauty, as for their fairness and soundness. Indeed, I have no faith in the selected lists of pomological gentlemen. Their "Favorites" and "None-suches" and "Seek-no-farthers," when I have fruited them, commonly turn out very tame and forgettable. They are eaten with comparatively little zest, and have no real *tang* nor *smack* to them.

What if some of these wildings are acrid and puckery, genuine *verjuice*, do they not still belong to the *Pomaceæ*, which are uniformly innocent and kind to our race? I still begrudge them to the cider-mill. Perhaps they are not fairly ripe yet.

No wonder that these small and high-colored apples are thought to make the best cider. Loudon quotes from the "Herefordshire Report," that "apples of a small size are always, if equal in quality, to be preferred to those of a larger size, in order that the rind and kernel may bear the greatest proportion to the pulp, which affords the weakest and most watery juice." And he says that, "to prove this, Dr. Symonds, of Hereford, about the year 1800, made one hogshead of cider entirely from the rinds and cores of apples, and another from the pulp only, when the first was found of extraordinary strength and flavor, while the latter was sweet and insipid."

Evelyn says that the "Red-strake" was the favorite cider-apple in his day; and he quotes one Dr. Newburg as saying, "In Jersey 't is a general observation, as I hear, that the more of red any apple has in its rind, the more proper it is for this use. Pale-faced apples they exclude as much as may be from their cider-vat." This opinion still prevails.

All apples are good in November. Those which the farmer leaves out as unsalable and unpalatable to those who frequent the markets are choicest fruit to the walker. But it is remarkable that the wild apple, which I praise as so spirited and racy when eaten in the fields or woods, being brought into the house has frequently a harsh and crabbed taste. The Saunterer's Apple not even the saunterer can eat in the house. The palate rejects it there, as it does haws and acorns, and demands a tamed one; for there you miss the November air, which is the sauce it is to be eaten with. Accordingly, when Tityrus, seeing the lengthening shadows, invites Melibœus to go home and pass the night with him, he promises him *mild* apples and soft chestnuts, — *mitia poma, castaneæ molles*. I frequently pluck wild apples of so rich and spicy a flavor that I wonder all orchardists do not get a scion from that tree, and I fail not to bring home my pockets full. But perchance, when I take one out of my desk and taste it in my chamber, I find it unexpectedly crude, — sour enough to set a squirrel's teeth on edge and make a jay scream.

These apples have hung in the wind and frost and rain till they have absorbed the qualities of the weather or season, and thus are highly *seasoned*, and they *pierce* and *sting* and *permeate* us with their spirit. They must be eaten in *season*, accordingly, — that is, out-of-doors.

To appreciate the wild and sharp flavors of these October fruits, it is necessary that you be breathing the sharp October or November air. The outdoor air and exercise which the walker gets give a different tone to his palate, and he craves a fruit which the sedentary would call harsh and crabbed. They must be eaten in the fields, when your system is all aglow with exercise, when the frosty weather nips your fingers, the wind rattles the bare boughs or rustles the few remaining leaves, and the jay is heard screaming around. What is sour in the house a bracing walk makes sweet. Some of these apples might be labeled, "To be eaten in the wind."

Of course no flavors are thrown away; they are intended for the taste that is up to them. Some apples have two distinct flavors, and perhaps one half of them must be eaten in the house, the other outdoors. One Peter Whitney wrote from Northborough in 1782, for the Proceedings of the Boston Academy, describing an apple tree in that town "producing fruit of opposite qualities, part of the same apple being frequently sour and the other sweet;" also some all sour, and others all sweet, and this diversity on all parts of the tree.

There is a wild apple on Nawshawtuct Hill in my town which has to me a peculiarly pleasant bitter tang, not perceived till it is three-quarters tasted. It remains on the tongue. As you eat it, it smells exactly like a squash-bug. It is a sort of triumph to eat and relish it.

I hear that the fruit of a kind of plum tree in Provence is "called *Prunes sibarelles*, because it is impossible to whistle after having eaten them, from their sourness."

But perhaps they were only eaten in the house and in summer, and if tried out-of-doors in a stinging atmosphere, who knows but you could whistle an octave higher and clearer?

In the fields only are the sours and bitters of Nature appreciated; just as the woodchopper eats his meal in a sunny glade, in the middle of a winter day, with content, basks in a sunny ray there, and dreams of summer in a degree of cold which, experienced in a chamber, would make a student miserable. They who are at work abroad are not cold, but rather it is they who sit shivering in houses. As with temperatures, so with flavors; as with cold and heat, so with sour and sweet. This natural raciness, the sours and bitters which the diseased palate refuses, are the true condiments.

Let your condiments be in the condition of your senses. To appreciate the flavor of these wild apples requires vigorous and healthy senses, *papillæ* firm and erect on the tongue and palate, not easily flattened and tamed.

From my experience with wild apples, I can understand that there may be reason for a savage's preferring many kinds of food which the civilized man rejects. The former has the palate of an outdoor man. It takes a savage or wild taste to appreciate a wild fruit.

What a healthy out-of-door appetite it takes to relish the apple of life, the apple of the world, then!

> "Nor is it every apple I desire,
> Nor that which pleases every palate best;
> 'T is not the lasting Deuxan I require,
> Nor yet the red-cheeked Greening I request,

> Nor that which first beshrewed the name of wife,
> Nor that whose beauty caused the golden strife:
> No, no! bring me an apple from the tree of life."

So there is one *thought* for the field, another for the house. I would have my thoughts, like wild apples, to be food for walkers, and will not warrant them to be palatable if tasted in the house.

THEIR BEAUTY

Almost all wild apples are handsome. They cannot be too gnarly and crabbed and rusty to look at. The gnarliest will have some redeeming traits even to the eye. You will discover some evening redness dashed or sprinkled on some protuberance or in some cavity. It is rare that the summer lets an apple go without streaking or spotting it on some part of its sphere. It will have some red stains, commemorating the mornings and evenings it has witnessed; some dark and rusty blotches, in memory of the clouds and foggy, mildewy days that have passed over it; and a spacious field of green reflecting the general face of nature, — green even as the fields; or a yellow ground, which implies a milder flavor, — yellow as the harvest, or russet as the hills.

Apples, these I mean, unspeakably fair, — apples not of Discord, but of Concord! Yet not so rare but that the homeliest may have a share. Painted by the frosts, some a uniform clear bright yellow, or red, or crimson, as if their spheres had regularly revolved, and enjoyed the influence of the sun on all sides alike, — some with the faintest pink blush imaginable, — some brindled with deep red streaks like a cow, or with hundreds of fine blood-red rays running regularly from the stem-dimple to the blossom end, like meridional lines, on a straw-colored ground, — some touched with a greenish rust, like a fine lichen, here and there, with crimson blotches or eyes more or less confluent and fiery when wet, — and others gnarly, and freckled or peppered all over on the stem side with fine crimson spots on a white ground, as if accidentally sprinkled from the brush of Him who paints the autumn leaves. Others, again, are sometimes red inside, perfused with a beautiful blush, fairy food, too beautiful to eat, — apple of the Hesperides, apple of the evening sky! But like shells and pebbles on the sea-shore, they must be seen as they sparkle amid the withering leaves in some dell in the woods, in the autumnal air, or as they lie in the wet grass, and not when they have wilted and faded in the house.

THE NAMING OF THEM

It would be a pleasant pastime to find suitable names for the hundred varieties which go to a single heap at the cider-mill. Would it not tax a man's invention, — no one to be named after a man, and all in the *lingua vernacula?* Who shall stand godfather at the christening of the wild apples? It would exhaust the Latin and Greek languages, if they were used, and make the *lingua vernacula* flag. We should have to call in the sunrise and the sunset, the rainbow and the autumn woods and the wild-flowers, and the woodpecker and the purple finch and the squirrel and the jay and the butterfly, the November traveler and the truant boy, to our aid.

In 1836 there were in the garden of the London Horticultural Society more than fourteen hundred distinct sorts. But here are species which they have not in their catalogue, not to mention the varieties which our crab might yield to cultivation.

Let us enumerate a few of these. I find myself compelled, after all, to give the Latin names of some for the benefit of those who live where English is not spoken, — for they are likely to have a world-wide reputation.

There is, first of all, the Wood Apple (*Malus sylvatica*); the Blue-Jay Apple; the Apple which grows in Dells in the Woods (*sylvestrivallis*), also in Hollows in Pastures (*campestrivallis*); the Apple that grows in an old Cellar-Hole (*Malus cellaris*); the Meadow Apple; the Partridge Apple; the Truant's Apple (*cessatoris*), which no boy will ever go by without knocking off some, however *late* it may be; the Saunterer's Apple, — you must lose yourself before you can find the way to that; the Beauty of the Air (*decus aëris*); December-Eating; the Frozen-Thawed (*gelato-soluta*), good only in that state; the Concord Apple, possibly the same with the *Musketaquidensis;* the Assabet Apple; the Brindled Apple; Wine of New England; the Chickaree Apple; the Green Apple (*Malus viridis*), — this has many synonyms: in an imperfect state, it is the *choleramorbifera aut dysenterifera, puerulis dilectissima;* the Apple which Atalanta stopped to pick up; the Hedge Apple (*Malus sepium*); the Slug Apple (*limacea*); the Railroad Apple, which perhaps came from a core thrown out of the cars; the Apple whose Fruit we tasted in our Youth; our Particular Apple, not to be found in any catalogue; *pedestrium solatium;* also the Apple where hangs the Forgotten Scythe; Iduna's Apples, and the Apples which Loki found in the Wood; and a great many more I have on my list, too numerous to mention, — all of them good. As Bodæus exclaims, referring to the cultivated kinds, and adapting Virgil to his case, so I, adapting Bodæus, —

> "Not if I had a hundred tongues, a hundred mouths,
> An iron voice, could I describe all the forms
> And reckon up all the names of these *wild apples*."

THE LAST GLEANING

By the middle of November the wild apples have lost some of their brilliancy, and have chiefly fallen. A great part are decayed on the ground, and the sound ones are more palatable than before. The note of the chickadee sounds now more distinct, as you wander amid the old trees, and the autumnal dandelion is half closed and tearful. But still, if you are a skillful gleaner, you may get many a pocketful even of grafted fruit, long after apples are supposed to be gone out-of-doors. I know a Blue Pearmain tree, growing within the edge of a swamp, almost as good as wild. You would not suppose that there was any fruit left there, on the first survey, but you must look according to system. Those which lie exposed are quite brown and rotten now, or perchance a few still show one blooming cheek here and there amid the wet leaves. Nevertheless, with experienced eyes, I explore amid the bare alders and the huckleberry bushes and the withered sedge, and in the crevices of the rocks, which are full of leaves, and pry under the fallen and decaying ferns, which, with apple and alder leaves, thickly strew the ground. For I know that they lie concealed, fallen into hollows long since and covered up by the leaves of the tree itself, — a proper kind of packing. From these lurking-places, anywhere within the circumference of the tree, I draw forth the fruit, all wet and glossy, maybe nibbled by rabbits and hollowed out by crickets, and perhaps with a leaf or two cemented to it (as Curzon an old manuscript from a monastery's mouldy cellar), but still with a rich bloom on it, and at least as ripe and well-kept, if not better than those in barrels, more crisp and lively than they. If these resources fail to yield anything, I have learned to look between the bases of the suckers which spring thickly from some horizontal limb, for now and then one lodges there, or in the very midst of an alder-clump, where they are covered by leaves, safe from cows which may have smelled them out. If I am sharp-set, for I do not refuse the Blue Pearmain, I fill my pockets on each side; and as I retrace my steps in the frosty eve, being perhaps four or five miles from home, I eat one first from this side, and then from that, to keep my balance.

I learn from Topsell's Gesner, whose authority appears to be Albertus, that the following is the way in which the hedgehog collects and carries home his apples. He says, — "His meat is apples, worms, or grapes: when he findeth apples or grapes on the earth, he rolleth himself upon them, until he have filled all his prickles, and then carrieth them home to his den, never bearing above one in his mouth; and if it fortune that one of them fall off by the way, he likewise shaketh off all the residue, and walloweth upon them afresh, until they be all settled upon his back again. So, forth he goeth, making a noise like a cart-wheel; and if he have any young ones in his nest, they pull off his load wherewithal he is loaded, eating thereof what they please, and laying up the residue for the time to come."

THE "FROZEN-THAWED" APPLE

Toward the end of November, though some of the sound ones are yet more mellow and perhaps more edible, they have generally, like the leaves, lost their beauty, and are beginning to freeze. It is finger-cold, and prudent farmers get in their barreled apples, and bring you the apples and cider which they have engaged; for it is time to put them into the cellar. Perhaps a few on the ground show their red cheeks above the early snow, and occasionally some even preserve their color and soundness under the snow throughout the winter. But generally at the beginning of the winter they freeze hard, and soon, though undecayed, acquire the color of a baked apple.

Before the end of December, generally, they experience their first thawing. Those which a month ago were sour, crabbed, and quite unpalatable to the civilized taste, such at least as were frozen while sound, let a warmer sun come to thaw them, — for they are extremely sensitive to its rays, — are found to be filled with a rich, sweet cider, better than any bottled cider that I know of, and with which I am better acquainted than with wine. All apples are good in this state, and your jaws are the cider-press. Others, which have more substance, are a sweet and luscious food, — in my opinion of more worth than the pineapples which are imported from the West Indies. Those which lately even I tasted only to repent of it, — for I am semicivilized, — which the farmer willingly left on the tree, I am now glad to find have the property of hanging on like the leaves of the young oaks. It is a way to keep cider sweet without boiling. Let the frost come to freeze them first, solid as stones, and then the rain or a warm winter day to thaw them, and they will seem to have borrowed a flavor from heaven through the medium of the air in which they hang. Or perchance you find, when you get home, that those which rattled in your pocket have thawed, and the ice is turned to cider. But after the third or fourth freezing and thawing they will not be found so good.

What are the imported half-ripe fruits of the torrid south, to this fruit matured by the cold of the frigid north? These are those crabbed apples with which I cheated my companion, and kept a smooth face that I might tempt him to eat. Now we both greedily fill our pockets with them, — bending to drink the cup and save our lappets from the overflowing juice, — and grow more social with their wine. Was there one that hung so high and sheltered by the tangled branches that our sticks could not dislodge it?

It is a fruit never carried to market, that I am aware of, — quite distinct from the apple of the markets, as from dried apple and cider, — and it is not every winter that produces it in perfection.

The era of the Wild Apple will soon be past. It is a fruit which will probably become extinct in New England. You may still wander through old orchards of native fruit of great extent, which for the most part went to the cider-mill, now all gone to decay. I have heard of an orchard in a distant town, on the side of a hill, where the apples rolled down and lay four feet deep against a wall on the lower side, and this the owner cut down for fear they should be made into cider. Since the temperance reform and the general introduction of grafted fruit, no native apple trees, such as I see everywhere in deserted pastures, and where the woods have grown up around them, are set out. I fear that he who walks over these fields a century hence will not know the pleasure of knocking off wild apples. Ah, poor man, there are many pleasures which he will not know! Notwithstanding the prevalence of the Baldwin and the Porter, I doubt if so extensive orchards are set out to-day in my town as there were a century ago, when those vast straggling cider-orchards were planted, when men both ate and drank apples, when the pomace-heap was the only nursery, and trees cost nothing but the trouble of setting them out. Men could afford then to stick a tree by every wall-side and let it take its chance. I see nobody planting trees to-day in such out of the way places, along the lonely roads and lanes, and at the bottom of dells in the wood. Now that they have grafted trees, and pay a price for them, they collect them into a plat by their houses, and fence them in, — and the end of it all will be that we shall be compelled to look for our apples in a barrel.

This is "The word of the Lord that came to Joel the son of Pethuel.

"Hear this, ye old men, and give ear, all ye inhabitants of the land! Hath this been in your days, or even in the days of your fathers? . . .

"That which the palmerworm hath left hath the locust eaten; and that which the locust hath left hath the cankerworm eaten; and that which the cankerworm hath left hath the caterpillar eaten.

"Awake, ye drunkards, and weep; and howl, all ye drinkers of wine, because of the new wine; for it is cut off from your mouth.

"For a nation is come up upon my land, strong, and without number, whose teeth are the teeth of a lion, and he hath the cheek teeth of a great lion.

"He hath laid my vine waste, and barked my fig tree: he hath made it clean bare, and cast it away; the branches thereof are made white. . . .

"Be ye ashamed, O ye husbandmen; howl, O ye vinedressers. . . .

"The vine is dried up, and the fig tree languisheth; the pomegranate tree, the palm tree also, and the apple tree, even all the trees of the field, are withered: because joy is withered away from the sons of men."

HUCKLEBERRIES

Agrestem tenui meditabor arundine musam
I am going to play a rustic strain on my
slender reed —
non injussa cano —
but I trust that I do not sing unbidden things.

MANY public speakers are accustomed, as I think foolishly, to talk about what they call *little* things in a patronising way sometimes, advising, perhaps, that they be not wholly neglected; but in making this distinction they really use no juster measure than a ten-foot pole, and their own ignorance. According to this rule a small potatoe is a little thing, a big one a great thing. A hogshead-full of anything — the big cheese which it took so many oxen to draw — a national salute — a state-muster — a fat ox — the horse Columbus — or Mr. Blank — the Oinan Boy — there is no danger that any body will call these *little things*. A cartwheel is a great thing — a snow flake a little thing. The *Wellingtonia gigantea* — the famous California tree, is a great thing — the seed

*The ornamental device (§) in the text marks those places where Professor Leo Stoller determined the order of the text by internal evidence—usually Thoreau's notes to himself about placement of the mancscript. For a description of the manuscripts and the principles which governed their editing, see Stoller's *Huckleberries* (The Windhover Press of the University of Iowa and The New York Public Library, 1970).

from which it sprang a little thing — scarcely one traveller has noticed the seed at all — and so with all the seeds or origins of things. But Pliny said — *In minimis Natura praestat* — Nature excels in the least things.

In this country a political speech, whether by Mr. Seward or Caleb Cushing, is a great thing, a ray of light a little thing. It would be felt to be a greater national calamity if you should take six inches from the corporeal bulk of one or two gentlemen in Congress, than if you should take a yard from their wisdom and manhood.

I have noticed that whatever is thought to be covered by the word *education* — whether reading, writing or 'rithmetick — is a great thing, but almost all that constitutes education is a little thing in the estimation of such speakers as I refer to. In short, whatever they know and care but *little* about is a little thing, and accordingly almost everything good or great is little in their sense, and is very slow to grow any bigger.

When the husk gets separated from the kernel, almost all men run after the husk and pay their respects to that. It is only the husk of Christianity that is so bruited and wide spread in this world, the kernel is still the very least and rarest of all things. There is not a single church founded on it. To obey the higher law is generally considered the last manifestation of littleness.

I have observed that many English naturalists have a pitiful habit of speaking of their proper pursuit as a sort of trifling or waste of time — a mere interruption to more important employments and 'severer studies' — for which they must ask pardon of the reader. As if they would have you believe that all the rest of their lives was consecrated to some truly great and serious enterprise. But it happens that we never hear more of this, as we certainly should, if it were only some great public or philanthropic service, and therefore conclude that they have been engaged in the heroic and magnanimous enterprise of feeding, clothing, housing and warming themselves and their dependents, the chief value of all which was that it enabled them to pursue just these studies of which they speak so slightingly. The 'severer study' they refer to was keeping their accounts. Comparatively speaking — what they call their graver pursuits and severer studies was the real trifling and misspense of life — and were they such fools as not to know it? It is, in effect at least, mere cant. All mankind have depended on them for this intellectual food.

I presume that every one of my audience knows what a huckleberry is — has seen a huckleberry — gathered a huckleberry — nay tasted a huckleberry — and that being the case, that you will not be averse to revisiting the huckleberry field in imagination this evening, though the pleasure of this excursion may fall as far short of the reality, as the flavor of a dried huckleberry is inferior to that of a fresh one. §*Huckleberries* begin to be ripe July third (or generally the thirteenth), are thick enough to pick about the twenty-second, at their height about the fifth of August, and last fresh till after the middle of that month.

This, as you know, is an upright shrub more or less stout depending on the exposure, with a spreading bushy top — a dark brown bark — red recent shoots and thick leaves. The flowers are smaller and much more red than those of the other species.

It is said to range from the Saskatchewan to the mountains of Georgia, and from the Atlantic to the Mississippi in this latitude — but it abounds over but a small part of this area, and there are large tracts where it is not found at all.

By botanists it is called of late, but I think without good reason, *Gaylussacia resinosa*, after the celebrated French chemist. If he had been the first to distil its juices and put them in this globular bag, he would deserve this honor, or if he had been a celebrated picker of huckleberries, perchance paid for his schooling so, or only notoriously a lover of them, we should not so much object. But it does not appear that he ever saw one. What if a committee of Parisian naturalists had been appointed to break this important news to an Indian maiden who had just filled her basket on the shore of Lake Huron! It is as if we should hear that the Daguerreotype had been *finally* named after the distinguished Chippeway conjurer, The-Wind-that-Blows. By another it has been called *Andromeda baccata*, the berry bearing andromeda — but he evidently lived far away from huckleberries and milk.

I observe green huckleberries by the nineteenth of June, and perhaps three weeks later, when I have forgotten them, I first notice on some hill side exposed to the light, some black or blue ones amid the green ones and the leaves, always sooner than I had expected, and though they may be manifestly premature, I make it a point to taste them, and so inaugurate the huckleberry season.

In a day or two the black are so thick among the green ones that they no longer incur the suspicion of being worm-eaten, and perhaps a day later I pluck a handful from one bush, and I do not fail to make report of it when I get home, though it is rarely believed, most people are so behind hand in their year's accounts.

Early in August, in a favorable year, the hills are black with them. At Nagog Pond I have seen a hundred bushels in one field — the bushes drooping over the rocks with the weight of them — and a very handsome sight they are, though you should not pluck one of them. They are of various forms, colors and flavors — some round — some pear-shaped — some glossy black — some dull black, some blue with a tough and thick skin (though they are never of the peculiar light blue of blueberries with a bloom) — some sweeter, some more insipid — etc., etc., more varieties than botanists take notice of.

To-day perhaps you gather some of those large, often pear-shaped, sweet blue ones which grow tall and thinly amid the rubbish where woods have been cut. They have not borne there before for a century

— being over-shadowed and stinted by the forest — but they have the more concentrated their juices — and profited by the new recipes which nature has given them — and now they offer to you fruit of the very finest flavor, like wine of the oldest vintage.

And tomorrow you come to a strong moist soil where the black ones shine with such a gloss — every one its eye on you, and the blue are so large and firm, that you can hardly believe them to be huckleberries at all, or edible; but you seem to have travelled into a foreign country, or else are dreaming.

They are a firmer berry than most of the whortleberry family — and hence are the most marketable.

If you look closely at a huckleberry you will see that it is dotted, as if sprinkled over with a yellow dust or meal, which looks as if it could be rubbed off. Through a microscope, it looks like a resin which has exuded, and on the small green fruit is of a conspicuous light orange or lemon color, like small specks of yellow lichens. It is apparently the same with that shining resinous matter which so conspicuously covers the leaves when they are unfolding, making them sticky to the touch — whence this species is called *resinosa* or resinous.

§There is a variety growing in swamps — a very tall and slender bush drooping or bent like grass to one side — commonly three or four feet high, but often seven feet — the berries, which are later than the former, are round and glossy black — with resinous dots as usual — and grow in flattish topped racemes — sometimes ten or twelve together, though generally more scattered. I call it the swamp-huckleberry.

But the most marked variety is the *red*-huckleberry — the *white* of some, (for the less ripe are whitish) — which ripens at the same time with the black. It is red with a white cheek, often slightly pear-shaped, semitransparent with a luster, very finely and indistinctly white dotted. It is as easily distinguished from the common in the green state as when ripe. I know of but three or four places in the town where they grow. It might be called *Gaylussacia resinosa var. erythrocarpa.*

I once did some surveying for a man, who remarked, but not till the job was nearly done, that he did not know when he should pay me. I did not at first pay much heed to this observation, though it was unusual, supposing that he meant to pay me within a reasonable time. Nevertheless it occurred to me that if he did not know when he should pay me still less did I know when I should be paid. He added, however, that I was perfectly secure, for there were the pigs in the stye (and as nice pigs as ever were seen) and there was his farm itself which I had surveyed, and knew was there as much as he. All this had its due influence in increasing my sense of security, as you may suppose. After many months he sent me a quart of red huckleberries, for they grew on his farm, and this I thought was ominous; he distinguished me altogether too much by this gift, since I was not his particular friend. I saw that it was the first installment of my dues — and that it would go a great way toward being the last. In the course of years he paid a part of the debt in money, and that is the last that I have heard of it. I shall beware of red huckleberry gifts in the future.

§Then there is the Late Whortleberry — Dangleberry or Blue Tangles — whose fruit does not begin to be ripe until about a month after huckleberries begin, when these and blueberries are commonly shrivelled and spoiling — on about August seventh, and is in its prime near the end of August.

This is a tall and handsome bush about twice as high as a huckleberry bush, with altogether a glaucous aspect, growing in shady copses where it is rather moist, and to produce much fruit it seems to require wet weather.

The fruit is one of the handsomest of berries, smooth, round and blue, larger than most huckleberries and more transparent, on long stems dangling two or three inches, and more or less tangled. By the inexperienced it is suspected to be poisonous, and so avoided, and perhaps is the more fair and memorable to them on that account. Though quite good to eat, it has a peculiar, slightly astringent, and compared with most huckleberries, not altogether pleasant flavor, and a tough skin.

At the end of the first week of September, they are commonly the only edible Whortleberries which are quite fresh. They are rare hereabouts however, and it is only in certain years that you can find enough for a pudding.

There is still another kind of Huckleberry growing in this town, called the *Hairy Huckleberry*, which ripens about the same time with the last. It is quite rare, growing only in the wildest and most neglected places, such as cold sphagneous swamps where the *Andromeda polifolia* and *Kalmia glauca* are found, and in some almost equally neglected but firmer low ground. The berries are oblong and black, and, with us, roughened with short hairs. It is the only species of *Vaccinieae* that I know of in this town whose fruit is inedible; though I have seen another kind of Whortleberry, the *Deer-berry* or Squaw Huckleberry, growing in another part of the state, whose fruit is said to be equally inedible. The former is merely insipid however. Some which grow on firmer ground have a little more flavor, but the thick and shaggy-feeling coats of the berries left in the mouth are far from agreeable to the palate.

Both these and Dangleberries are placed in the same genus (or section) with the common huckleberry.

Huckleberries are very apt to dry up and not attain their proper size — unless rain comes to save them before the end of July. They will be dried quite hard and black by drought even before they have ripened. On the other hand they frequently burst open and are so spoiled in consequence of copious rains when they are fully ripe.

§They *begin* to be soft and wormy as early as the middle of August, and generally about the twentieth the children cease to carry them round to sell, as they are suspected by the purchasers.

How late when the huckleberries begin to be wormy

and the pickers are deserting the fields! The walker feels very solitary now.

But in woods and other cool places they commonly last quite fresh a week or more longer, depending on the season. In some years when there are far more berries than pickers or even worms, and the birds appear to pass them by, I have found them plump, fresh, and quite thick, though with a somewhat dried taste, the fourteenth of October, when the bushes were mostly leafless, and the leaves that were left, were all red, and they continued to hold on after the leaves had all fallen, till they were softened and spoiled by rain.

Sometimes they begin to dry up generally by the middle of August — after they are ripe, but before spoiling, and by the end of that month I have seen the bushes so withered and brown owing to the drought, that they appeared dead like those which you see broken off by the pickers, or as if burnt.

I have seen the hills still black with them, though hard and shrivelled as if dried in a pan, late in September. And one year I saw an abundance of them still holding on the eleventh of December, they having dried ripe prematurely, but these had no sweetness left. The sight of them thus dried by nature may have originally suggested to the Indians to dry them artificially.

High-blueberries, the second kind of low-blueberries, huckleberries and low-blackberries are all at their height generally during the first week of August. In the dog-days (or the first ten of them) they abound and attain their full size.

Huckleberries are classed by botanists with the cranberries (both bog and mountain) — snowberry, bearberry — mayflower, checkerberry — the andromedas, clethra, laurels, azaleas, rhodora, ledum, pyrolas, prince's pine, Indian pipes, and many other plants, and they are called all together the Heath Family, they being in many respects similar to and occupying similar ground with the heaths of the Old World, which we have not. If the first botanists had been American this might have been called the Huckleberry Family including the heaths. Plants of this order (*Ericaceae*) are said to be among the earliest ones found in a fossil state, and one would say that they promised to last as long as any on this globe. George B. Emerson says that the whortleberry differs from the heath proper, 'essentially only in its juicy fruit surrounded by the calyx segments.'

The genus to which the whortleberries belong, is called by most botanists *Vaccinium*, which I am inclined to think is properly derived from *bacca*, a berry, as if these were the chief of all berries, though the etymology of this word is in dispute.

Whortle- or Hurtleberry, Bilberry, and Blae or Blea, that is *blue* berry are the names given in England originally to the fruit of the *Vaccinium myrtillus* which we have not in New England and also to the more scarce and local *Vaccinium uliginosum* which we have.

The word whortleberry is said to be derived from the Saxon *heortberg* (or *heorot-berg*), the hart's berry.

Hurts is an old English word used in heraldry, where, according to Bailey, it is 'certain balls resembling hurtleberries.'

The Germans say *Heidel-berre* — that is *heath berry.*

Huckleberry — this word is used by Lawson in 1709 — appears to be an American word derived from Whortleberry — and applied to fruits of the same family, but for the most part of different species from the English whortleberries. According to the Dictionary the word berry is from the Saxon *beria* — a grape or cluster of grapes. A French name of whortleberry is 'raisin des bois' — grape of the woods. It is evident that the word berry has a new significance in America.

We do not realize how rich our country is in berries. The ancient Greeks and Romans appear not to have made much account of strawberries, huckleberries, melons etc. because they had not got them.

The Englishman Lindley, in his *Natural System of Botany,* says that the *Vaccinieae* are 'Natives of North America, where they are found in great abundance as far as high northern latitudes; sparingly in Europe; and not uncommonly on high land in the Sandwich Islands.'

Or as George B. Emerson states it, they 'are found chiefly in the temperate, or on the mountains in the warmer regions of America. Some are found in Europe; some on the continent and islands of Asia, and on islands in the Atlantic, Pacific and Indian Oceans.' 'The whortleberries and cranberries,' says he, 'take the place, throughout the northern part of this continent, of the heaths of the corresponding climates of Europe; and fill it with no less of beauty, and incomparably more of use.'

According to the last arrangement of our plants, we have fourteen species of the Whortleberry Family (*Vaccinieae*) in New England, eleven of which bear edible berries, eight, berries which are eaten raw, and five of the last kind are abundant — to wit — the huckleberry — the bluet or Pennsylvania blueberry — the Canada blueberry (in the northern part of New England) — the second or common low blueberry — and the high or swamp-blueberry (not to mention the Dangleberry, which is common in some seasons and localities).

On the other hand I gather from London and others that there are only two species growing in England, which are eaten raw, answering to our eight — to wit, the Bilberry (*V. myrtillus*) and the Blea-berry or Bog Whortleberry (*V. uliginosum*), both of which are found in North America, and the last is the common one on the summit of the White Mountains, but in Great Britain it is found only in the northern part of England and in Scotland. This leaves only one in England to our five which are abundant.

In short, it chances that of the thirty-two species of *Vaccinium* which Loudon describes, all except the above two and four more are referred to North America alone, and only three or possibly four are found in Europe.

Yet the few Englishmen with whom I have spoken on this subject love to think and to say that they have as many huckleberries as we. I will therefore quote the most which their own authorities say not already quoted, about the abundance and value of their only two kinds which are eaten raw.

Loudon says of the bog whortleberry (*V. uliginosum*), 'The berries are agreeable but inferior in flavor to those of *Vaccinium myrtillus* [the bilberry]; eaten in large quantities, they occasion giddiness, and a slight headache.'

And of their common whortleberry (*V. myrtillus*) he says, 'It is found in every country in Britain, from Cornwall to Caithness, least frequently in the south-eastern countries, and increases in quantity as we advance northward.' It 'is an elegant and also a fruit-bearing plant.' The berries 'are eaten in tarts or with cream, or made into a jelly, in the northern and western counties of England; and, in other parts of the country they are made into pies and puddings.' They 'are very acceptable to children either eaten by themselves, or with milk' or otherwise. They 'have an astringent quality.'

Coleman in his "Woodlands, Heaths, and Hedges," says 'The traveller in our upland and mountain districts can hardly have failed to notice, as his almost constant companion, this cheerful little shrub, . . . it fluorishes best in a high airy situation, only the summits of the very loftiest mountains of which this country can boast being too elevated for this hardy little mountaineer.' 'In Yorkshire, and many parts of the north, large quantities of bilberries are brought into the market, being extensively used as an ingredient in pies and puddings, or preserved in the form of jam. . . . Much, however, of the relish of these wilding fruits must be set down to the exhilirating air, and those charms of scenery that form the accessories of a mountain feast; . . . One of the prettiest sights that greet one's eye in the districts where it abounds, is that of a party of rustic children "a-bilberrying" (for the greater portion of those that come to market are collected by children); there they may be seen, knee deep in the "wires," or clambering over the broken gray rocks to some rich nest of berries, their tanned faces glowing with health, and their picturesque dress (or undress) — with here and there bits of bright red, blue, or white — to the painter's eye contrasting beautifully with the purple, gray and brown of the moorland, and forming altogether rich pictorial subjects.'

These authorities tell us that children and others eat the fruit, just as they tell us that the birds do. It is evident from all this that whortleberries do not make an important part of the regular food of the Old English people in their season, as they do of the New Englanders. What should we think of a summer in which we did not taste a huckleberry pudding? That is to Jonathan what his plum pudding is to John Bull.

Yet Dr. Manassah Cutler, one of the earliest New England botanists, speaks of the huckleberry lightly as being merely a fruit which children love to eat with their milk. What ingratitude thus to shield himself behind the children! I should not wonder if it turned out that Dr. Manassah Cutler ate his huckleberry pudding or pie regularly through the season, as many his equals do. I should have pardoned him had he frankly put in his thumb and pulled out a plum, and cried 'What a Great Doctor am I?' But probably he was lead astray by reading English books or it may be that the Whites did not make so much use of them in his time.

Widely dispersed as their bilberry may still be in England, it was undoubtedly far more abundant there once. One botanist says that 'This is one of the species, that if allowed, would overrun Britain, and form, with *Culluna vulgaris* (heather) and *Empetrum nigrum* (crowberry, which grows on our White Mountains), much of the natural physiognomical character of its vegetation.'

The genus *Gaylussacia*, to which our huckleberry belongs, has no representative in Great Britain, nor does our species extend very far northward in this country.

So I might say of edible berries generally, that there are far fewer kinds in Old than in New England.

Take the *rubuses* or what you might call bramble berries, for instance, to which genus our raspberries, blackberries and thimbleberries belong. According to Loudon there are five kinds indigenous in Britain to our eight. But of these five only two appear to be at all common, while we have four kinds both very common and very good. The Englishman Coleman says of their best, the English raspberry, which species we also cultivate, that 'the wilding is not sufficiently abundant to have much importance.'

And the same is true of wild fruits generally. Hips and haws are much more important comparatively there than here, where they have hardly got any popular name.

I state this to show how contented and thankful we ought to be.

It is to be remembered that the vegetation in Great Britain is that of a much more northern latitude than where we live, that some of our alpine shrubs are found on the plain there and their two whortleberries are alpine or extreme northern plants with us.

If you look closely you will find blueberry and huckleberry bushes under your feet, though they may be feeble and barren, throughout all our woods, the most persevering Native Americans, ready to shoot up into place and power at the next election among the plants, ready to reclothe the hills when man has laid them bare and feed all kinds of pensioners. What though the woods be cut down; it appears that this emergency was long ago anticipated and provided for by Nature, and the interregnum is not allowed to be a barren one. She not only begins instantly to heal that scar, but she compensates us for the loss and refreshes us with fruits such as the forest did not produce. As the sandal wood is said to diffuse a perfume around the woodman who cuts it — so in this case Nature rewards with unexpected fruits the hand that lays her waste.

I have only to remember each year where the woods have been cut just long enough to know where to look for them. It is to refresh us thus once in a century that they bide their time on the forest floor. If the farmer mows and burns over his overgrown pasture for the benefit of grass, or to keep the children out, the huckleberries spring up there more vigorous than ever, and the fresh blueberry shoots tinge the earth crimson. All our hills are, or have been, huckleberry hills, the three hills of Boston and no doubt Bunker Hill among the rest. My mother remembers a woman who went a-whortleberrying where Dr. Lowell's church now stands.

In short the whortleberry bushes in the Northern States and British America are a sort of miniature forest surviving under the great forest, and reappearing when the latter is cut, and also extending northward beyond it. The small berry-bearing shrubs of this family, as the crowberry, bilberry, and cranberry, are called by the Esquimaux in Greenland, 'berry grass,' and Crantz says that the Greenlanders cover their winter houses with 'bilberry bushes,' together with turf and earth. They also burn them; and I hear that somebody in this neighborhood has invented a machine for cutting up huckleberry bushes for fuel.

It is remarkable how universally, as it respects soil and exposure, the whortleberry family is distributed with us, almost we may say a new species for every thousand feet of elevation. One kind or another, of those of which I am speaking, fluorishing in every soil and locality.

There is the high blueberry in swamps — the second low blueberry, with the huckleberry, on almost all fields and hills — the Pennsylvanian and Canada blueberries especially in cool and airy places in openings in the woods and on hills and mountains, while we have two kinds confined to the alpine tops of our highest mountains — the family thus ranging from the lowest valleys to the highest mountain tops, and forming the prevailing small shrubbery of a great part of New England.

The same is true *hereabouts* of a single species of this family, the huckleberry proper. I do not know of a spot where any shrub grows in this neighborhood, but one or another variety of the huckleberry may also grow there. It is stated in Loudon that all the plants of this order 'require a peat soil, or a soil of a close cohesive nature,' but this is not the case with the huckleberry. It grows on the tops of our highest hills — no pasture is too rocky or barren for it — it grows in such deserts as we have, standing in pure sand — and at the same time it flourishes in the strongest and most fertile soil. One variety is peculiar to quaking bogs where there can hardly be said to be any soil beneath, to say nothing of another but unpalatable species, the hairy huckleberry, which is found there. It also extends through all our woods more or less thinly, and a distinct species, the dangleberry, belongs especially to moist woods and thickets.

Such care has Nature taken to furnish to birds and quadrupeds, and to men, a palatable berry of this kind, slightly modified by soil and climate, wherever the consumer may chance to be. Corn and potatoes, apples and pears, have comparatively a narrow range, but we can fill our basket with whortleberries on the summit of Mount Washington, above almost all other shrubs with which we are familiar, the same kind which they have in Greenland, and again when we get home, with another species in our lowest swamps, such as the Greenlanders never dreamed of.

The berries *which I celebrate*, appear to have a range, most of them, very nearly coterminous with what has been called the Algonquin Family of Indians, whose territories embraced what are now the Eastern, Middle and Northwestern States — and the Canadas — and surrounded those of the Iroquois in what is now New York. These were the small fruits of the Algonquin and Iroquois Families.

Of course the Indians naturally made a much greater account of wild fruits than we do, and among the most important of these were huckleberries.

They taught us not only the use of corn and how to plant it, but also of whortleberries and how to dry them for winter. We should have hesitated long before we tasted some kinds if they had not set us the example, knowing by old experience that they were not only harmless but salutary. I have added a few to my number of edible berries, by walking behind an Indian in Maine, and observing that he ate some which I never thought of tasting before.

To convince you of the extensive use which the Indians made of huckleberries, I will quote at length the testimony of the most observing travellers, on this subject, as nearly as possible in the order in which it was given us; for it is only after listening patiently to such reiterated and concurring testimony, of various dates — and respecting widely distant localities — that we come to realize the truth.

But little is said by the discoverers of the use which the Indians made of the fresh berries in their season — the hand to mouth use of them — because there was little to be said — though in this form they may have been much the most important to them. We have volumes of recipes, called cookbooks — but when a fruit or a tart is ready for the table, nothing remains but to eat it without any more words. We therefore have few or no accounts of Indians going a-huckleberrying — though they had more than a six weeks' vacation for that purpose, and probably camped on the huckleberry field.

I will go far enough back for my authorities to show that they did not learn the use of these berries from us whites.

In the year 1615, Champlain, the founder of Quebec, being far up the Ottawa spying out the land and taking notes among the Algonquins, on his way to the Fresh Water Sea since called Lake Huron — observed that the natives made a business of collecting and drying for winter use, a small berry which he called blues, and also raspberries — the former is the common blueberry of those regions, by some considered a variety of our early low blueberry (*Vaccinium Pennsylvanicum*); and again when near the lake he observes that the natives make a kind of bread of pounded corn sifted and mixed with mashed beans which have been boiled — and sometimes they put dried blueberries and raspberries into it.

This was five years before the Pilgrims crossed the Atlantic, and is the first account of huckleberry cake that I know of.

Gabriel Sagard, a Franciscan Friar, in the account of his visit to the Huron Country in 1624, says, 'There is so great a quantity of blues, which the Hurons call *Ohentaque*, and other little fruits which they call by a general name *Hahique*, that the savages regularly dry them for the winter, as we do prunes in the sun, and that serves them for comfits for the sick, and to give taste to their *sagamite* [or gruel, making a kind of plum porridge], and also to put into the little loaves (or cakes, *pains*) which they cook under the ashes.'

According to him they put not only blueberries and raspberries into their bread but strawberries, 'wild mulberries (*meures champestres*) and other little fruits dry and green.'

Indeed the gathering of blueberries by the savages is spoken of by the early French explorers as a regular and important harvest with them.

LeJeune, the Superior of the Jesuits in Canada — residing at Quebec — in his Relation for 1639 — says of the savages that 'Some figure to themselves a paradise full of *bluets*.'

Roger Williams, who knew the Indians well, in his account of those in his neighborhood — published in 1643 — tells us that '*Sautaash* are those currants (grapes and whortleberries) dried by the natives, and so preserved all the year, which they beat to powder and mingle it with their parched meal, and make a delicate dish which they call *Sautauthig*, which is as sweet to them as plum or spice cake to the English.'

But Nathaniel Morton, in his *New England's Memorial*, printed in 1669 — speaking of white men going to treat with Canonicus, a Narraghanset Indian, about Mr. Oldham's death in 1636 — says 'Boiled chestnuts is their white bread, and because they would be extraordinary in their feasting, they strove for variety after the English manner, boiling puddings made of beaten corn, putting therein great store of blackberries, somewhat like currants' — no doubt whortleberries. This *seems* to *imply* that the Indians imitated the English — or set before their guests dishes to which they themselves were not accustomed — or which were extra-ordinary. But we have seen that these dishes were not new or unusual to them and it was the whites who imitated the Indians rather.

John Josselyn — in his *New England Rarities*, published in 1672 — says under the fruits of New England, 'Bill-berries, two kinds, black and sky colored, which is more frequent. . . . The Indians dry them in the sun and sell them to the English by the bushel, who make use of them instead of currence, putting of them into puddens, both boyled and baked, and into water gruel.'

The largest Indian huckleberry party that I have heard of is mentioned in the life of Captain Church who, it is said, when in pursuit of King Phillip in the summer of 1676, came across a large body of Indians, chiefly squaws, gathering whortleberries on a plain near where New Bedford now is, and killed and took prisoner sixty-six of them — some throwing away their baskets and their berries in their flight. They told him that their husbands and brothers, a hundred of them, who with others had their rendezvous in a great cedar swamp nearby, had recently left them to gather whortleberries there, while they went to Sconticut Neck to kill cattle and horses for further and more substantial provisions.

La Hontan in 1689, writing from the Great Lakes, repeats what so many French travellers had said about the Indians drying and preserving blueberries — saying, 'The savages of the north make a great harvest of them in summer, which is a great resource especially when the chase fails them.' They were herein more provident than we commonly suppose.

Father Raslles — who was making a Dictionary of the Abenaki Language in 1691 (at Norridgewock?) — says that their word for blueberries was fresh *Satar*, dry *Sakisatar* — and the words in their name for July meant when the blueberries are ripe. This shows how important they were to them.

Father Hennepin — who writes in 1697 — says that his captors, Naudowessi (the Sioux!), near the falls of St. Anthony, feasted on wild-rice seasoned with blueberries, 'which they dry in the sun during the summer, and which are as good as raisins of Corinth' — [that is, the imported currants].

The Englishman John Lawson, who published an account of the Carolinas in 1709, says of North Carolina, 'The hurts, huckleberries or blues of this country are four sorts. . . . The first sort is the same blue or bilberry that grows plentifully in the North of England.' 'The second sort grows on a small bush,' the fruit being larger than the last. The third grows three or four feet high in low land. 'The fourth sort grows upon trees, some ten and twelve foot high, and the thickness of a man's arm; these are found in the runs and low grounds. . . . The Indians get many bushels, and dry them on mats, whereof they make plum bread, and many other eatables.' He is the first author that I remember who uses the word 'huckleberry.'

The well known natural botanist John Bartram, when returning to Philadelphia in 1743 from a Journey through what was then the wilderness of Pennsylvania and New York, to the Iroquois and Lake Ontario, says that he 'found [when in Pennsylvania] an Indian squaw drying huckleberries. This is done by setting four forked sticks in the ground, about three or four feet high, then others across, over them the stalks of our common *Jacea* or *Saratula*, on these lie the berries, as malt is spread on the hair cloth over the kiln. Underneath she had kindled a smoke fire, which one of her children was tending.'

Kalm, in his travels in this country in 1748-9,

writes, 'On my travels through the country of the Iroquois, they offered me, whenever they designed to treat me well, fresh maize bread, baked in an oblong shape, mixed with dried huckleberries, which lay as close in it as the raisins in a plumb pudding.'

The Moravian missionary Heckewelder, who spent a great part of his life among the Delawares toward the end of the last century, states that they mixed with their bread, which was six inches in diameter by one inch thick — 'whortleberries green or dry, but not boiled.'

Lewis and Clarke in 1805 found the Indians west of the Rocky Mountains using dried berries extensively.

And finally in Owen's Geological Survey of *Wisconsin, Iowa and Minnesota* — published in 1852 — occurs the following. '*Vaccinium Pennslvanicum* (*Lam.*) [that is, our early low blueberry] — Barrens on the upper St. Croix. This is the common Huckleberry, associated with the characteristic growth of the *Pinus Banksiana*, covering its sandy ridges with a verdant undergrowth, and an unsurpassed luxuriance of fruit. By the Indians these are collected and smoke dried in great quantities, and in this form constitute an agreeable article of food.'

Hence you see that the Indians from time immemorial, down to the present day, all over the northern part of America — have made far more extensive use of the whortleberry — at all seasons and in various ways — than we — and that they were far more important to them than to us.

It appears from the above evidence that the Indians used their dried berries commonly in the form of a cake, and also of huckleberry porridge or pudding.

What we call huckleberry cake made of Indian meal and huckleberries was evidently the principal cake of the aborigines — though they also used other berries and fruits in a similar manner and often put things into their cake which would not have been agreeable to our palates — though I do not hear that they ever put any soda or pearl ash or alum into it. We have no national cake so universal and well known as this was in all parts of the country where corn and huckleberries grew.

They enjoyed it all alone ages before our ancestors heard of their Indian corn — or their huckleberries — and probably if you had travelled here a thousand years ago it would have been offered you alike on the Connecticut, the Potomac, the Niagara, the Ottawa and the Mississippi.

The last Indian of Nantucket, who died a few years ago, was very properly represented in a painting which I saw there, with a basket full of huckleberries in his hand, as if to hint at the employment of his last days. I trust that I may not outlive the last of the huckleberries.

Tanner, who was taken captive by the Indians in 1789, and spent a good part of his life as an Indian, gives the Chippeway names of at least five kinds of whortleberries. He gives '*meen* — blue berry, *meen-un* — blue berries,' and says that 'this is a word that enters into the composition of almost all words which are used as the names of fruits,' that is as a terminal syllable. Hence this would appear to have been the typical berry — or berry of berries — among the Chippeway as it is among us.

I think that it would be well if the Indian names, were as far as possible restored and applied to the numerous species of huckleberries, by our botanists — instead of the very inadequate — Greek and Latin or English ones at present used. They might serve both a scientific and popular use. Certainly it is not the best point of view to look at this peculiarly American family as it were from the other side of the Atlantic. It is still in doubt whether the Latin word for the genus *Vaccinium* means a berry or a flower.

Botanists, on the look out for what they thought a respectable descent, have long been inclined to trace this family backward to Mount Ida. Tourneforte does not hestitate to give it the ancient name of *Vine of Mount Ida.* The common English Raspberry also is called *Rubus Idaea* or the Mount Ida bramble — from the old Greek name. The truth of it seems to be that blueberries and raspberries flourish best in cool and airy situations, on hills and mountains, and I can easily believe that something *like* these at least grows on Mount Ida. But Mount Monadnoc is as good as Mount Ida, and probably better for blueberries, though its name is said to mean *Bad Rock.* But the worst rocks are the best for poets' uses. Let us then exchange that oriental uncertainty for this western certainty.

We have in the northern states a few wild plums

and inedible crab apples — a few palatable grapes — and many tolerable nuts — but I think that the various species of berries are our *wild fruits* which are to be compared with the more celebrated ones of the tropics, and for my part I would not exchange fruits with them — for the object is not merely to get a ship-load of something which you can eat or sell, but the pleasure of gathering it is to be taken into the account.

What is the pear crop as yet to the huckleberry crop? Horticulturists make a great ado about their pears, but how many families raise or buy a barrel of pears in a year all told? They are comparatively insignificant. I do not taste more than half a dozen pears annually, and I suspect that the majority fare worse even than I. (This was written before my neighbor's pear-orchard began to bear. Now he frequently fills my own and others' pockets with the fruit.) But Nature heaps the table with berries for six weeks or more. Indeed the apple crop is not so important as the huckleberry crop. Probably the apples consumed in this town annually do not amount to more than one barrel per family. But what is this to a month or more of huckleberrying to every man, woman and child, and the birds into the bargain. Even the crop of oranges, lemons, nuts, raisins, figs, quinces, etc, is of little importance to us compared with these.

They are not unprofitable in a pecuniary sense; I hear that some of the inhabitants of Ashby sold $2000 worth of huckleberries in '56.

In May and June all our hills and fields are adorned with a profusion of the pretty little more or less bell-shaped flowers of this family, commonly turned toward the earth and more or less tinged with red or pink, and resounding with the hum of insects, each one the forerunner of a berry the most natural, wholesome and palatable that the soil can produce. I think to myself, these are the blossoms of the *Vaccinieae* or Whortleberry family, which affords so large a portion of our berries; the berry-promising flower of the *Vaccinieae*! This crop grows wild all over the country — wholesome, bountiful and free, a real ambrosia. And yet men, the foolish demons that they are, devote themselves to the culture of tobacco, inventing slavery and a thousand other curses for that purpose — with infinite pains and inhumanity go raise tobacco all their lives, and that is the staple instead of huckleberries. Wreathes of tobacco smoke go up from this land, the only incense which its inhabitants burn in honor of their gods. With what authority can such as we distinguish between Christians and Mahometans? Almost every interest, as the codfish and mackerel interest, gets represented at the General Court — but not the huckleberry interest. The first discoverers and explorers of the land make report of this fruit — but the last make comparatively little account of them.

Blueberries and huckleberries are such simple, wholesome and universal fruits that they concern our race much. It is hard to imagine any country without this kind of berry, on which men live like birds.

still covering our hills as when the red men lived here. Are they not the principal wild fruit?

What means this profusion of berries at this season only? Nature does her best to feed her children, and the broods of birds just matured find plenty to eat now. Every bush and vine does its part and offers a wholesome and palatable diet to the way-farer. He need not go out of the road to get as many berries as he wants — of various kinds and qualities according as his road leads him over high or low, wooded or open ground — huckleberries of different colors and flavors almost everywhere — the second kind of low blueberry largest in the moist ground — high blueberries with their agreeable acid when his way lies through a swamp, and low blackberries of two or more varieties on almost every sandy plain and bank and stone heap.

Man at length stands in such a relation to Nature as the animals which pluck and eat as they go. The fields and hills are a table constantly spread. Diet-drinks, cordials, wines of all kinds and qualities, are bottled up in the skins of countless berries for their refreshment, and they quaff them at every turn. They seem offered to us not so much for food as for sociality, inviting us to a pic-nic with Nature. We pluck and eat in remembrance of her. It is a sort of sacrament — a communion — the *not* forbidden fruits, which no serpent tempts us to eat. Slight and innocent savors which relate us to Nature, make us her guests, and entitle us to her regard and protection.

When I see, as now, in climbing one of our hills,

huckleberry and blueberry bushes bent to the ground with fruit, I think of them as fruits fit to grow on the most Olympian or heaven-pointing hills.

It does not occur to you at first that where such thoughts are suggested is Mount Olympus, and that you who taste these berries are a god. Why in his only royal moments should man abdicate his throne?

You eat these berries in the dry pastures where they grow not to gratify an appetite, but as simply and naturally as thoughts come into your mind, as if they were the food of thought, dry as itself, and surely they nourish the brain there.

Occasionally there is an unusual profusion of these fruits to compensate for the scarcity of a previous year. I remember some seasons when favorable moist weather had expanded the berries to their full size, so that the hill-sides were literally black with them. There were infinitely more of all kinds than any and all creatures could use.

One such year, on the side of Conantum Hill, they were literally five or six inches deep. First, if you searched low down in the shade under all, you found still fresh the great light blue earliest blueberries, bluets, in heavy clusters — that most Olympian fruit of all — delicate flavored, thin-skinned and cool — then, next above, the still denser masses or clusters of the second low blueberry of various varieties, firm and sweet food — and rising above these large blue and black huckleberries of various qualities — and over these ran rampant the low blackberry weighing down the thicket with its wreathes of black fruit, and binding it together in a trembling mass — while here and there the high blackberry, just beginning to be ripe, towered over all the rest. Thus, as it were, the berries hung up lightly in masses or heaps, separated by their leaves and twigs so that the air could circulate through and preserve them; and you went daintily wading through this thicket, picking perhaps only the finest of the high blackberries, as big as the end of your thumb, however big that may be, or clutching here and there a handful of huckleberries for variety, but never suspecting the delicious, cool blue-bloomed ones, which you were crushing with your feet under all. I have in such a case spread aside the bushes and revealed the last kind to those who had never in all their lives seen or heard of it before.

Each such patch, each bush — seems fuller and blacker than the last, as you proceed, and the huckleberries at length swell so big, as if aping the blackberries, that you mark the spot for future years.

There is all this profusion and yet you see neither birds nor beasts eating them — only ants and the huckleberry-bug. It seems fortunate for us that those cows in their pasture do not love them, but pass them by. We do not perceive that birds and quadrupeds make any use of them because they are so abundant we do not miss them, and they are not compelled to come when we are for them. Yet they are far more important to them than to us. We do not notice the robin when it plucks a huckleberry as we do when it visits our favorite cherry tree — and the fox pays his visits to the fields when we are not there.

I once carried my arms full of these bushes to my boat, and while I was rowing homeward two ladies, who were my companions, picked three pints from these alone, casting the bare bushes into the stream from time to time.

Even in ordinary years, when berries are comparatively scarce, I sometimes unexpectedly find so many in some distance and unfrequented part of the town, between and about the careless farmers' houses and walls, that the soil seems more fertile than where I live. Every bush and bramble bears its fruit. The very sides of the road are a fruit garden. The earth there teems with blackberries, huckleberries, thimbleberries, fresh and abundant — no signs of drought nor of pickers. Great shining black berries peep out at me from under the leaves upon the rocks. Do the rocks hold moisture? or are there no fingers to pluck these fruits? I seem to have wandered into a land of greater fertility — some up country Eden. These are the Delectable Hills. It is land flowing with milk and huckleberries, only they have not yet put the berries into the milk. *There* the herbage never withers, *there* are abundant dews. I ask myself, What are the virtues of the inhabitants that they are thus blessed?

A fortunatos nimium, sua si bona norint Agricolas—

O too fortunate husbandmen if they knew their own happiness.

These berries are further important as introducing children to the fields and woods. The season of berrying is so far respected that the school children have a vacation then — and many little fingers are busy picking these small fruits. It is even a pastime, not a drudgery — though it often pays well beside. The First of August is to them the anniversary of Emancipation in New England.

Women and children who never visit distant hills, fields and swamps on any other errand are seen making haste thither now with half their domestic utensils in their hands. The wood-chopper goes into the swamp for fuel in the winter; his wife and children for berries in the summer.

Now you will see who is the thorough countrywoman who does not go to the beach — conversant with berries and nuts — a masculine wild-eyed woman of the fields.

Now for a ride in the hay-rigging to that far off Elysium that Zachariah See-all alighted on — but has not mentioned to every person — in the hay-rigging without springs — trying to sensitive nerves and to full pails, for all alike sit on the bottom — such a ride is favorable to conversation for the incessant rumble hides all defects and fills the otherwise aweful pauses — to be introduced to new scenes more memorable than the berries— but to the old walker the straggling party itself half concealed amid the bushes is the most novel and interesting feature. If hot the boys break up the bushes and carry them to some shady place where the girls can pick them at their ease. But this is a lazy and improvident way — and gives an unsightly look to the hill. There are many events not in the program. If you have an ear for music — perhaps one is the sound of a cow bell — never heard before — or a sudden thunder shower putting you to flight — or a breakdown.

I served my apprenticeship and have since done considerable journeywork in the huckleberry field. Though I never paid for my schooling and clothing in that way, it was some of the best schooling that I got and paid for itself. Theodore Parker is not the only New England boy who has got his education by picking huckleberries, though he may not have gone to Harvard thereafter, nor to any school more distant than the huckleberry field. *There* was the university itself where you could learn the everlasting Laws, and Medicine and Theology, not under Story, and Warren, and Ware, but far wiser professors than they. Why such haste to go from the huckleberry field to the College yard?

As in old times they who dwelt on the heath, remote from towns, being backward to adopt the doctrines which prevailed in towns, were called heathen in a bad sense, so I trust that we dwellers in the huckleberry pastures, which are our heath-lands, shall be slow to adopt the notions of large towns and cities, though perchance we may be nicknamed huckleberry people. But the worst of it is that the emissaries of the towns come more for our berries than they do for our salvation.

Occasionally, in still summer forenoons, when perhaps a mantua-maker was to be dined, and a huckleberry pudding had been decided on (by the authorities), I a lad of ten was despatched to a neighboring hill alone. My scholastic education could be thus far tampered with, and an excuse might be found. No matter how scarce the berries on the near hills, the exact number necessary for a pudding could surely be collected by eleven o'clock — and all ripe ones too though I turned some round three times to be sure they were not premature. My rule in such cases was never to eat one till my dish was full; for going a-berrying implies more things than eating the berries. They at home got nothing but the pudding, a comparatively heavy affair — but I got the forenoon out of doors — to say nothing about the appetite for the pudding. They got only the plums that were in the pudding, but I got the far sweeter plums that never go into it.

At other times, when I had companions, some of them used to bring such remarkably shaped dishes, that I was often curious to see how the berries disposed of themselves in them. Some brought a coffee-pot to the huckleberry field, and such a vessel possessed this advantage at least, that if a greedy boy had skimmed off a handful or two on his way home, he had only to close the lid and give his vessel a shake to have it full again. I have seen this done all round when the party got as far homeward as the Dutch House. It can probably be done with any vessel that has much side to it.

There was a Young America then, which has become Old America, but its principles and motives are still the same, only applied to other things. Sometimes, just before reaching the spot — every boy rushed to the hill side and hastily selecting a spot — shouted 'I speak for this place,' indicating its bounds, and another 'I speak for that,' and so on — and this was sometimes considered good law for the huckleberry field. At any rate it is a law similar to this by which we have taken possession of the territory of Indians and Mexicans.

I once met with a whole family, father, mother, and children, ravaging a huckleberry field in this wise. They cut up the bushes as they went and beat them over the edge of a bushel basket, till they had it full of berries, ripe and green, leaves, sticks etc., and so they passed along out of my sight like wild men.

I well remember with what a sense of freedom and spirit of adventure I used to take my way across the fields with my pail, some years later, toward some distant hill or swamp, when dismissed for all day, and I would not now exchange such an expansion of all my being for all the learning in the world. Liberation and enlargement — such is the fruit which all culture aims to secure. I suddenly knew more about my books than if I had never ceased studying them. I found myself in a schoolroom where I could not fail to see and hear things worth seeing and hearing — where I could not help getting my lesson — for my lesson came to me. Such experience often repeated, was the chief encouragement to go to the Academy and study a book at last.

But ah we have fallen on evil days! I hear of pickers ordered out of the huckleberry fields, and I see stakes set up with written notices forbidding any to pick them. Some let their fields or allow so much for the picking. *Sic transit gloria ruris.* I do not mean to blame any, but all —to bewail our fates generally. We are not grateful enough that we have lived a part of our lives before these things occurred. What becomes of the true value of country life — what, if you must go to market for it? It has come to this, that the butcher now brings round our huckleberries in his cart. Why, it is as if the hangman were to perform the marriage ceremony. Such is the inevitable tendency of our civilization, to reduce huckleberries to a level with beef-steaks — that is to blot out four fifths of it, or the going a-huckleberrying, and leave only a pudding, that part which is the fittest accompaniment to a beef-steak. You all know what it is to go a-beef-steaking. It is to knock your old fellow laborer Bright on the head to begin with — or possibly to cut a steak from him running in the Abyssinian fashion and wait for another to grow there. The butcher's item in chalk on the door is now 'Calf's head and huckleberries.'

I suspect that the inhabitants of England and the continent of Europe have thus lost in a measure their natural rights, with the increase of population and monopolies. The wild fruits of the earth disappear before civilization, or only the husks of them are to be found in large markets. The whole country becomes, as it were, a town or beaten common, and almost the only fruits left are a few hips and haws.

What sort of a country is that where the huckleberry fields are private property? When I pass such fields on the highway, my heart sinks within me. I see a blight on the land. Nature is under a veil there. I make haste away from the accursed spot. Nothing could deform her fair face more. I cannot think of it ever after but as the place where fair and palatable berries, are converted into money, where the huckleberry is desecrated.

It is true, we have as good a right to make berries private property, as to make wild grass and trees such — it is not worse than a thousand other practices which custom has sanctioned — but that is the worst of it, for it suggests how bad the rest are, and to what result our civilization and division of labor natually tend, to make all things venal.

A., a professional huckleberry picker, has hired B.'s field, and, we will suppose, is now gathering the crop, with a patent huckleberry horse rake.

C., a professed cook, is superintending the boiling of a pudding made of some of the berries.

While Professor D. — for whom the pudding is intended, sits in his library writing a book — a work on the *Vaccinieae* of course.

And now the result of this downward course will be seen in that work — which should be the ultimate fruit of the huckleberry field. It will be worthless. It will have none of the spirit of the huckleberry in it, and the reading of it will be a weariness of the flesh.

I believe in a different kind of division of labor — that Professor D. should be encouraged to divide himself freely between his library and the huckleberry field.

What I chiefly regret in this case, is the in effect dog-in-the-manger result; for at the same time that we exclude mankind from gathering berries in our field, we exclude them from gathering health and happiness and inspiration, and a hundred other far finer and nobler fruits than berries, which are found there, but which we have no notion of gathering and shall not gather ourselves, nor ever carry to market, for these is no market for them, but let them rot on the bushes.

We thus strike only one more blow at a simple and wholesome relation to nature. I do not know but this is the excuse of those who have lately taken to swinging bags of beans and ringing dumb bells. As long as the berries are free to all comers they are beautiful, though they may be few and small, but tell me that this is a blueberry swamp which somebody has hired, and I shall not want even to look at it. We so commit the berries to the wrong hands, that is to the hands of those who cannot appreciate them. This is proved by the fact that if we do not pay them some money, these parties will at once cease to pick them. They have no other interest in berries but a pecuniary one. Such is the constitution of our society that we make a compromise and permit the berries to be degraded, to be enslaved, as it were.

Accordingly in laying claim for the first time to the spontaneous fruit of our pastures, we are inevitably aware of a little meanness, and the merry berry party which we turn away naturally looks down on and despises us. If it were left to the berries to say who should have them, is it not likely that they would prefer to be gathered by the party of children in the hay-rigging, who have come to have a good time merely?

This is one of the taxes which we pay for having a rail-road. All our improvements, so called, tend to convert the country into the town. But I do not see clearly that these successive losses are ever quite made up to us. This suggests, as I have said, what origin and foundation many of our institutions have. I do not say this by way of complaining of this custom in particular, which is beginning to prevail — not that I love Caesar less but Rome more. It is my own way of living that I complain of as well as yours — and therefore I trust that my remarks will come home to you. I hope that I am not so poor a shot, like most clergymen, as to fire into a crowd of a thousand men without hitting somebody — though I do not aim at any one.

Thus we behave like oxen in a flower garden. The true fruit of Nature can only be plucked with a fluttering heart and a delicate hand, not bribed by any earthly reward. No hired man can help us to gather that crop.

Among the Indians, the earth and its productions generally were common and free to all the tribe, like the air and water — but among us who have supplanted the Indians, the public retain only a small yard or common in the middle of the village, with perhaps a grave-yard beside it, and the right of way, by sufferance, by a particular narrow route, which is annually becoming narrower, from one such yard to another. I doubt if you can ride out five miles in any direction without coming to where some individual is tolling in the road — and he expects the time when it will all revert to him or his heirs. This is the way we civilized men have arranged it.

I am not overflowing with respect and gratitude to the fathers who thus laid out our New England villages, whatever precedent they were influenced by, for I think that a 'prentice hand liberated from Old English prejudices could have done much better in this new world. If they were in earnest seeking thus far away 'freedom to worship God,' as some assure us — why did they not secure a little more of it, when it was so cheap and they were about it? At the same time that they built meeting-houses why did they not preserve from desecration and destruction far grander temples not made with hands?

What are the natural features which make a township handsome — and worth going far to dwell in? A river with its water-falls — meadows, lakes — hills, cliffs or individual rocks, a forest and single ancient trees — such things are beautiful. They have a high use which dollars and cents never represent. If the inhabitants of a town were wise they would seek to preserve these things though at a considerable expense. For such things educate far more than any hired teachers or preachers, or any at present recognized system of school education.

I do not think him fit to be the founder of a state or even of a town who does not foresee the use of these things, but legislates as it were, for oxen chiefly.

It would be worth the while if in each town there were a committee appointed, to see that the beauty of the town received no detriment. If here is the largest boulder in the country, then it should not belong to an individual nor be made into door-steps. In some countries precious metals belong to the crown — so here more precious objects of great natural beauty should belong to the public.

Let us try to keep the new world new, and while we make a wary use of the city, preserve as far as possible the advantages of living in the country.

I think of no natural feature which is a greater ornament and treasure to this town than the river. It is one of the things which determine whether a man will live here or in another place, and it is one of the first objects which we show to a stranger. In this respect we enjoy a great advantage over those neighboring towns which have no river. Yet the town, as a corporation, has never turned any but the most purely utilitarian eyes upon it — and has done nothing to preserve its natural beauty.

They who laid out the town should have made the river available as a common possession forever. The town collectively should at least have done as much as an individual of taste who owns an equal area commonly does in England. Indeed I think that not only the channel but one or both banks of every river should be a public highway — for a river is not useful merely to float on. In this case, one bank might have been reserved as a public walk and the trees that adorned it have been protected, and frequent avenues have been provided leading to it from the main street. This would have cost but a few acres of land and but little wood, and we should all have been gainers by it. Now it is accessible only at the bridges at points comparatively distant from the town, and there there is not a foot of shore to stand on unless you trespass on somebody's lot — and if you attempt a quiet stroll down the bank — you soon meet with fences built at right angles with the stream and projecting far over the water — where individuals, naturally enough, under the present arrangement — seek to monopolize the shore. At last we shall get our only view of the stream from the meeting house belfry.

As for the trees which fringed the shore within my remembrance — where are they? and where will the remnant of them be after ten years more?

So if there is any central and commanding hilltop, it should be reserved for the public use. Think of a mountain top in the township — even to the Indians a sacred place — only accessible through private grounds. A temple as it were which you cannot enter without trespassing — nay the temple itself private property and standing in a man's cow yard — for such is commonly the case. New Hampshire courts have lately been deciding, as if it was for them to decide, whether the top of Mount Washington belonged to A or B — and it being decided in favor of B, I hear that he went up one winter with the proper officers and took formal possession. That area should be left unappropriated for modesty and reverence's sake — if only to suggest that the traveller who climbs thither in a degree rises above himself, as well as his native valley, and leaves some of his grovelling habits behind.

I know it is a mere figure of speech to talk about temples nowadays, when men recognize none and associate the word with heathenism. Most men, it appears to me, do not care for Nature, and would sell their share in all her beauty, for as long as they may live, for a stated and not very large sum. Thank God they cannot yet fly and lay waste the sky as well as the earth. We are safe on that side for the present. It is for the very reason that some do not care for these things that we need to combine to protect all from the vandalism of a few.

It is true, we as yet take liberties and go across lots in most directions but we naturally take fewer and fewer liberties every year, as we meet with more resistance, and we shall soon be reduced to the same straights they are in England, where going across lots is out of the question — and we must ask leave to walk in some lady's park.

There are a few hopeful signs. There is the growing *library* — and then the town does set trees along the highways. But does not the broad landscape itself deserve attention?

We cut down the few old oaks which witnessed the transfer of the township from the Indian to the white man, and perchance commence our museum with a cartridge box taken from a British soldier in 1775.§ How little we insist on truly grand and beautiful natural features. There may be the most beautiful landscapes in the world within a dozen miles of us, for aught we know — for their inhabitants do not value nor perceive them — and so have not made them known to others — but if a grain of gold were picked up there, or a pearl found in a fresh-water clam, the whole state would resound with the news.

Thousands annually seek the White Mountains to be refreshed by their wild and primitive beauty — but when the country was discovered a similar kind of beauty prevailed all over it — and much of this might have been preserved for our present refreshment if a little foresight and taste had been used.

I do not believe that there is a town in this country which realizes in what its true wealth consists.

I visited the town of Boxboro only eight miles west of us last fall — and far the handsomest and most memorable thing which I saw there, was its noble oak wood. I doubt if there is a finer one in Massachusetts. Let it stand fifty years longer and men will make pilgrimages to it from all parts of the country, and for a worthier object than to shoot squirrels in it — and yet I said to myself, Boxboro would be very like the rest of New England, if she were ashamed of that wood-land. Probably, if the history of this town is written, the historian will have omitted to say a word about this forest — the most interesting thing in it — and lay all the stress on the history of the parish.

It turned out that I was not far from right — for not long after I came across a very brief historical notice of Stow — which then included Boxboro —

written by the Reverend John Gardiner in the *Massachusetts Historical Collections*, nearly a hundred years ago. In which Mr. Gardiner, after telling us who was his predecessor in the ministry, and when he himself was settled, goes on to say, 'As for any remarkables, I am of mind there have been the fewest of any town of our standing in the Province. . . . I can't call to mind above one thing worthy of public notice, and that is the grave of Mr. John Green' who it appears, when in England, 'was made clerk of the exchequer' by Cromwell. 'Whether he was excluded the act of oblivion or not I cannot tell,' says Mr. Gardiner. At any rate he returned to New England and as Gardiner tells us 'lived and died, and lies buried in this place.'

I can assure Mr. Gardiner that he was not excluded from the act of oblivion.

It is true Boxboro was less peculiar for its woods at that date — but they were not less interesting absolutely.

I remember talking a few years ago with a young man who had undertaken to write the history of his native town — a wild and mountainous town far up country, whose very name suggested a hundred things to me, and I almost wished I had the task to do myself — so few of the original settlers had been driven out — and not a single clerk of the exchequer buried in it. But to my chagrin I found that the author was complaining of want of materials, and that the crowning fact of his story was that the town had been the residence of General C— and the family mansion was still standing.

§I have since heard, however, that Boxboro is content to have that forest stand, instead of the houses and farms that might supplant it — not because of its beauty — but because the land pays a much larger tax now than it would then.

Nevertheless it is likely to be cut off within a few years for ship-timber and the like. It is too precious to be thus disposed of. I think that it would be wise for the state to purchase and preserve a few such forests.

If the people of Massachusetts are ready to found a professorship of Natural History — so they must see the importance of preserving some portions of nature herself unimpaired.

I find that the rising generation in this town do not know what an oak or a pine is, having seen only inferior specimens. Shall we hire a man to lecture on botany, on oaks for instance, our noblest plants — while we permit others to cut down the few best specimens of these trees that are left? It is like teaching children Latin and Greek while we burn the books printed in those languages.

I think that each town should have a park, or rather a primitive forest, of five hundred or a thousand acres, either in one body or several — where a stick should never be cut for fuel — nor for the navy, nor to make wagons, but stand and decay for higher uses — a common possession forever, for instruction and recreation.

All Walden wood might have been reserved, with

Walden in the midst of it, and the Easterbrooks country, an uncultivated area of some four square miles in the north of the town, might have been our huckleberry field. If any owners of these tracts are about to leave the world without natural heirs who need or deserve to be specially remembered, they will do wisely to abandon the possession to all mankind, and not will them to some individual who perhaps has enough already — and so correct the error that was made when the town was laid out. As some give to Harvard College or another Institution, so one might give a forest or a huckleberry field to Concord. This town surely is an institution which deserves to be remembered. Forget the heathen in foreign parts, and remember the pagans and salvages here.

We hear of cow commons and ministerial lots, but we want *men* commons and *lay* lots as well. There is meadow and pasture and woodlot for the town's poor, why not a forest and huckleberry field for the town's rich?

We boast of our system of education, but why stop at schoolmasters and schoolhouses? § We are all schoolmasters and our schoolhouse is the universe. To attend chiefly to the desk or schoolhouse, while we neglect the scenery in which it is placed, is absurd. If we do not look out we shall find our fine schoolhouse standing in a cow yard at last.

It frequently happens that what the city prides itself on most is its park — those acres which require to be the least altered from their original condition.

Live in each season as it passes; breathe the air,

drink the drink, taste the fruit, and resign yourself to the influences of each. Let these be your only diet-drink and botanical medicines.

In August live on berries, not dried meats and pemmican as if you were on shipboard making your way through a waste ocean, or in the Darien Grounds, and so die of ship-fever and scurvy. Some will die of ship-fever and scurvy in an Illinois prairie, they lead such stifled and scurvy lives.

Be blown on by all the winds. Open all your pores and breathe in all the tides of nature, in all her streams and oceans, at all seasons. Miasma and infection are from within, not without. The invalid brought to the brink of the grave by an unnatural life, instead of imbibing the great influence that nature is — drinks only of the tea made of a particular herb — while he still continues his unnatural life — saves at the spile and wastes at the bung. He does not love nature or his life and so sickens and dies and no doctor can save him.

Grow green with spring — yellow and ripe with autumn. Drink of each season's influence as a vial, a true panacea of all remedies mixed for your especial use. The vials of summer never made a man sick, only those which he had stored in his cellar. Drink the wines not of your own but of nature's bottling — not kept in a goat- or pig-skin, but in the skins of a myriad fair berries.

Let Nature do your bottling, as also your pickling and preserving.

For all nature is doing her best each moment to

make us well. She exists for no other end. Do not resist her. With the least inclination to be well we should not be sick. Men have discovered, or think that they have discovered the salutariness of a few wild things only, and not of all nature. Why nature is but another name for health. Some men think that they are not well in Spring or Summer or Autumn or Winter, (if you will excuse the pun) it is only because they are not indeed *well,* that is fairly *in* those seasons.